Doing Data Analysis
with
SPSS 10.0

Robert H. Carver

Jane Gradwohl Nash

Stonehill College

Duxbury
Thomson Learning™

Australia • Canada • Mexico • Singapore • Spain • United Kingdom • United States

Sponsoring Editor: *Seema Atwal*
Marketing Team: *Samantha Cabaluna, Tom Ziolkowski*
Editorial Assistant: *Emily Davidson*
Production Coordinator: *Kelsey McGee*
Production Service: *Matrix Productions/*
  Merrill Peterson

Manuscript Editor: *Kathy Behler*
Permissions Editor: *Mary Kay Hancharick*
Cover Design: *Laurie Albrecht*
Cover Photo: *PhotoDisc*
Print Buyer: *Vena Dyer*
Cover Printing, Printing and Binding: *Webcom, Ltd.*

For more information about this or any other Duxbury product, contact:
DUXBURY
511 Forest Lodge Road
Pacific Grove, CA 93950 USA
www.duxbury.com
1-800-423-0563 Thomson Learning Academic Resource Center)

For permission to use material from this work, contact us by
Web: www.thomsonrights.com
fax: 1-800-730-2215
phone: 1-800-730-2214

Printed in Canada

10   9   8   7   6   5   4

Library of Congress Cataloging-in-Publication Data

Carver, Robert H.
  Doing Data Analysis with SPSS 10.0/Robert H. Carver, Jane Gradwohl Nash.
    p. cm.
  ISBN 0 534-37475-1
    1. SPSS (Computer file) 2. Social sciences—Statistical methods
  Computer programs I. Nash, Jane Gradwohl. II. Title.

HA32.C37 2000   00-036091
519.5'0285'5369—dc2100

For Donna, Sam, and Ben, who teach me daily.

RHC

For my parents, David and Hanna Gradwohl, who always let me ask why.

JGN

Contents

Preface

The Changing Environment of Statistics Education

In the past decade or so, educators have come to reconsider the best approach to teaching and learning in applied statistics courses. With the widespread availability of personal computers, advances in statistical software, and the near-universal application of quantitative methods in many professions, courses now emphasize statistical reasoning more than computational skill development. Questions of *how* have given way to more challenging questions of *why*, *when*, and *what?*

Simultaneously, undergraduates are increasingly comfortable with software, expecting to use computers in their work. Colleges are seeking ways to integrate information technology efficiently into coursework. The introductory statistics course is an ideal place to augment traditional out-of-class assignments with structured computer exercises.

The goal of this book is to supplement an introductory undergraduate statistics course with a comprehensive set of self-paced exercises. Students can work independently, learning the software skills outside of class, while coming to understand the underlying statistical concepts and techniques. Instructors can teach statistics and statistical reasoning, rather than algebra or software.

The Approach of This Book

The book reflects the changes described above in several ways. First, and most obviously, it provides some training in the use of a powerful software package to relieve students of computational drudgery.

Second, each session is designed to address a statistical issue or need, rather than to feature a particular command or menu in the software. Third, nearly all of the datasets in the book are real, reflecting a variety of disciplines. Fourth, the sessions follow a traditional sequence, making the book compatible with many texts. Finally, as each session leads the student through the techniques, it also includes thought-provoking questions and challenges, engaging the student in the processes of statistical reasoning. In designing the lab exercises, we kept four ideas in mind:

- *Statistical reasoning, not computation, is the goal of the course.* This manual asks students questions throughout, balancing software instruction with reflection on the meaning of results.

- *Students arrive in the course ready to learn statistical reasoning.* They need not slog all the way through descriptive techniques before encountering the concept of inference. The exercises invite students to think about inferences from the start, and the questions grow in sophistication as students master new material.

- *Exploration of real data is preferable to artificial datasets.* With the exception of the famous Anscombe regression dataset and a few simulations, all of the datasets are real. Some are very old and some are quite current, and they cover a wide range of substantive areas.

- *Statistical topics, rather than software features, should drive the design of each lab session.* Each lab session features several SPSS functions selected for their relevance to the statistical concept under consideration.

This book provides a rigorous but limited introduction to the software. The SPSS Base 10.0 system is rich in features and options; this book makes no attempt to "cover" the entire package. Instead, the level of coverage is commensurate with an introductory course. There may be many ways to perform a given task in SPSS; generally, we show one way. This book provides a "foot in the door." Interested students and other users can explore the software possibilities via the extensive Help system or other standard SPSS documentation.

Using This Book

We presume that this book is being used as a supplementary text in an introductory-level statistics course. If your courses are like ours (one in a psychology department, the other in a business department), class time is a scarce resource. Adding new material is always a balancing act. As such, supplementary readings and assignments must be carefully integrated. We suggest that instructors use the sessions in this book in four different ways, tailoring the approach throughout the term to meet the needs of the students and course.

- *In-class activity:* Part or all of some sessions might best be done together in class, with each student at a computer. The instructor can comment on particular points and can roam to offer assistance. This may be especially effective in the earliest sessions.
- *Stand-alone assignments*: In conjunction with a topic covered in the principal text, sessions can be assigned as independent out-of-class work, along with selected Moving On... questions. This is our most frequently-used approach. Students independently learn the software, re-enforce the statistical concepts, and come to class with questions about any difficulties they encountered in the lab session.
- *Preparation for text-based case or problem*: An instructor may wish to use a textbook case for a major assignment. The relevant session may prepare the class with the software skills needed to complete the case.
- *Independent projects*: Sessions may be assigned to prepare students to undertake an independent analysis project designed by the instructor. Many of the data files provided with the book contain additional variables that are never used within sessions. These variables may form the basis for original analyses or explorations.

Solutions are available to instructors for all Moving On... and bold-faced questions. Instructors should consult their Duxbury sales representatives for details.

The Data Files

As previously noted, each of the data files provided with this book contains real data, much of it downloaded from public sites on the World Wide Web. You can download all files from the Duxbury Press web site. Appendix A describes each file and its source, and provides detailed definitions of each variable.

The data files were chosen to represent a variety of interests and fields, and to illustrate specific statistical concepts or techniques. No doubt, each instructor will have some favorite datasets that can be used with these exercises. Most textbooks provide datasets as well. For some tips on converting other datasets for use with SPSS, see Appendix B.

Note on Software Versions

The examples in this manual are based on SPSS Base 10.0, running under Windows 95, Windows 98, or Windows NT. Users of earlier Windows versions or the Student version will notice only minor differences with the figures and instructions in this book, and in a few instances, will need to take an alternate approach. Adopters using SPSS Base 9.0 should assign Appendix C in lieu of Session 1.

To the Student

This book has two goals: to help you understand the concepts and techniques of statistical analysis, and to teach you how to use one particular tool—SPSS—to perform such analysis. It can supplement but not replace your primary textbook or your classroom time. To get the maximum benefit from the book, you should take your time and work carefully. Read through a session before you sit down at the computer. Each session should require no more than about 30 minutes of computer time; there's little need to rush through them.

We have included dialog box images for each new command. In those instances where you must drag a variable name into a box, you will notice that the name sometimes overhangs the box. Upon release of the mouse button, the name is truncated.

You'll often see questions interspersed through the computer instructions. These are intended to shift your focus from "getting answers" to thinking about what the answers mean, whether they make sense, whether they surprise or puzzle you, or how they relate to what you have been doing in class. Attend to these questions, even when you aren't sure of their purpose.

You may also notice that we have sometimes placed our names on a graph. We do this intermittently to call your attention to the practice; you should always place your name on the graphs you create.

Each lab ends with a section called **Moving On....** You should also respond to the numbered questions in that section, as assigned by your instructor. Questions in the Moving On... sections are designed to challenge you. Sometimes, it is quite obvious how to proceed with your analysis; sometimes, you will need to think a bit before you issue your first command. The goal is to get you to engage in statistical thinking, integrating what you have learned throughout your course. There is much more to doing data analysis than "getting the answer," and these questions provide an opportunity to do realistic analysis.

As noted earlier, SPSS is a large and very powerful software package, with many capabilities. Many of the features of the program are beyond the scope of an introductory course, and do not figure in these exercises. However, if you are curious or adventurous, you should explore the menus and Help system. You may find a quicker, more intuitive, or more interesting way to approach a problem.

Typographical Conventions

Throughout this manual, certain symbols and typefaces are used consistently. They are as follows:

🖱 **Menu ➤ Sub-menu ➤ Command** The mouse icon indicates an action you take at the computer, using the mouse or keyboard. The bold type lists menu selections for you to make.

Dialog box headings are in this typeface.

Dialog box choices, variable names, and items you should type appear in this typeface.

File names (e.g., Colleges) appear in this typeface.

> 💻 A box like this contains an instruction requiring special care or information about something that may work differently on your computer system.

Bold italics in the text indicate a question that you should answer as you write up your experiences in the lab.

Acknowledgments

Like most authors, we owe many debts of gratitude for this book. This project enjoyed the support of Stonehill College through the annual Summer Grants and the Stonehill Undergraduate Research Experience (SURE) programs. As the SURE scholar in the preparation of the book, Jason Boyd contributed in myriad ways, consistently doing reliable, thoughtful, and excellent work. He tested every session, prepared instructors' solutions, researched datasets, critiqued sessions from a student perspective, and tied up loose ends. His contributions and collegiality were invaluable.

Many colleagues and students suggested or provided datasets. Student contributors were Jennifer Axon, Stephanie Duggan, Debra Elliott, Tara O'Brien, Erin Ruell, and Benjamin White. A big thank you goes out to our students in Introduction to Statistics and Quantitative Analysis for Business for pilot-testing many of the sessions and for providing useful feedback about them.

We thank our Stonehill colleagues Ken Branco, Lincoln Craton, Roger Denome, Jim Kenneally, and Bonnie Klentz for suggesting or sharing data, and colleagues from other institutions who supported our work: Chris France, Roger Johnson, Stephen Nissenbaum, Mark Popovksy, and Alan Reifman. Thanks also to the many individuals and organizations granting permission to use published data for these sessions; they are all identified in Appendix A.

At Duxbury Press, we enjoyed the guidance and encouragement of Curt Hinrichs, Carolyn Crockett, Sarah Kaminskis, and Seema Atwal. Thanks also go to Paul Baum at California State University, Northridge and to Dennis Jowaisas at Oklahoma City University, two reviewers whose constructive suggestions have improved the quality of this book.

ෲ   ෲ   ෲ

Finally, we thank our families.

I want to thank my husband, Justin Nash, for his unwavering support of my professional work, and my daughters, Hanna Gradwohl Nash and Sara Gradwohl Nash, for providing an enjoyable distraction from this project.

JGN

The Carver home team has been fabulous, as always. To Donna, my partner and counsel; to Sam and Ben, my cheering section and assistants. Thanks for the time, space, and encouragement. Sometimes it *does* help to hear, "Dad, why are you writing another book?"

RHC

About the Authors

Robert H. Carver is Professor of Business Administration at Stonehill College in Easton, Massachusetts where his teaching has been recognized with the College's annual Excellence in Teaching award. In addition to Business Statistics, he teaches courses in information systems as well as business and society. He holds an A.B. from Amherst College and a Ph.D. in Public Policy from the University of Michigan. He is the author of *Doing Data Analysis with Minitab 12* (Duxbury Press), and his work has appeared in *Publius, The Journal of Statistics Education, PS: Political Science & Politics, Public Administration Review, Public Productivity Review,* and *The Journal of Consumer Marketing.*

Jane Gradwohl Nash is Associate Professor of Psychology at Stonehill College. She earned her B.A. from Grinnell College and her Ph.D. from Ohio University. She enjoys teaching courses in the areas of statistics, cognitive psychology, and developmental psychology. Her research interests are in the area of knowledge structure and knowledge change (learning). She is the author of articles that have appeared in the *Journal of Educational Psychology, Organizational Behavior and Human Decision Processes, Journal of Chemical Education, Research in the Teaching of English,* and *Written Communication.*

Session 1

A First Look at SPSS 10.0

Objectives

In this session, you will learn to do the following:

- Launch and exit SPSS
- Enter quantitative and qualitative data in a data file
- Create and print a graph
- Get Help
- Save your work to a disk

Launching SPSS

Before starting this session, you should know how to run a program within the Windows 95, 98, or Windows NT operating system. All the instructions in this manual presume basic familiarity with the Windows environment.

> 💻 Check with your instructor for specific instructions about running Windows 95/98/NT on your system. Your instructor will also tell you where to find SPSS.

Click and hold the left mouse button on the **⊞ Start** button at the lower left of your screen, and drag the cursor to select **Programs**. In the list, locate and choose **SPSS 10.0 for Windows**. Click and release the mouse button to launch the program. Because SPSS is a large program, you may have to wait a few moments before the program is ready for use.

On the next page is an image of the screen you will see when SPSS is ready. First you will see a menu dialog box listing several options; behind it is the *Data Editor* which is used to display the data that you will analyze using the program. Later you will encounter the *output Viewer window* which displays the results of your analysis. Each window has a unique purpose, to be made clear in due course. It's important at the outset to have a sense of what each window is about.

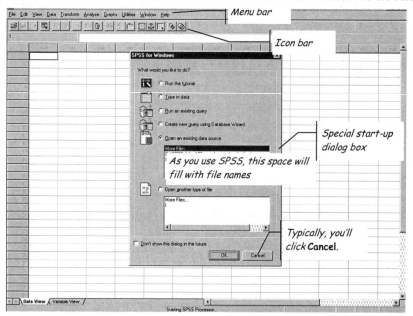

At any point in your session, only one window is *selected*, meaning that mouse actions and keystrokes will affect that window alone. When you start SPSS, the Data Editor is initially selected.

Since SPSS operates upon data, we generally start by placing data into the Editor, either from the keyboard or from a stored disk file. The Data Editor looks much like a spreadsheet. Cells may contain numbers or text, but unlike a spreadsheet, they never contain formulas. Except for the top row, which is reserved for variable names, rows are numbered consecutively. Each variable in your dataset will occupy one column of the SPSS data file, and each row represents one observation. For example, if you have a sample of fifty observations on two variables, your worksheet will contain two columns and fifty rows.

The menu bar across the top of the screen identifies broad categories of SPSS's features. There are two ways to issue commands in SPSS: choose commands from the menu or icon bars, or type them directly into a Syntax Editor. This book always refers you to the menus and icons. You can do no harm by clicking on a menu and reading the choices available, and you should expect to spend some time exploring your choices in this way.

Entering Data into the Data Editor

For most of the sessions in this book, you will start by accessing data already stored on a disk. For small datasets or class assignments, though, it will often make sense simply to type in the data yourself. For this session, you will transfer the data displayed below into the Data Editor.

In this first session, our goal is simple: to create a small data file, and then use SPSS to construct two graphs using the data. This is typical of the tasks you will perform throughout the book.

The coach of a high school swim team runs a practice for 10 swimmers, and records their times (in seconds) on a piece of paper.[1] Each swimmer is practicing the 50-meter freestyle event, and the boys on the team assert that they did better than the girls. The coach wants to analyze these results to see what the facts are. He codes gender with a 1 for the girls and a 2 for the boys.

Swimmer	Gender	Time
Sara	1	29.34
Jason	2	30.98
Joanna	1	29.78
Donna	1	34.16
Phil	2	39.66
Hanna	1	44.38
Sam	2	34.80
Ben	2	40.71
Abby	1	37.03
Justin	2	32.81

The first step in entering the data into the Data Editor is to define three variables: Swimmer, Gender, and Time. Creating a variable

---

[1] Nearly every dataset in this book is real. For the sake of starting modestly, we have taken a minor liberty in this session. This example is actually extracted from a dataset you will use later in the book. The full dataset appears in two forms: **Swimmer** and **Swimmer2**.

requires us to name it, specify the type of data (qualitative, quantitative, number of decimal places, etc.) and assign labels to the variable and data values if we wish.

🖱 Move your cursor to the bottom of the Data Editor, where you will see a tab labeled Variable View. Click on that tab. A different grid appears, with these column headings:

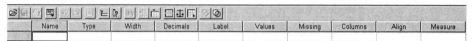

For each variable we create, we need to specify all or most of the attributes described by these column headings.

🖱 Move your cursor into the first empty cell in Row 1 (under **Name**) and type the variable name Swimmer. Press Enter (or Tab).

🖱 When the cursor moves to the **Type** column, a small gray button marked with three dots will appear; click on it and you'll see this dialog box. Numeric is the default variable type.

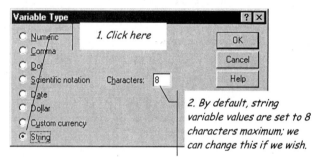

🖱 Click on the circle labeled String in the lower left corner of the dialog box. The names of the swimmers constitute a *nominal* or *categorical* variable, represented by a "string" of characters rather than a number. Click **Continue**.

Notice that the **Measure** column (far right column) now reads Nominal, because you chose String as the variable type.

In SPSS, each variable may carry a descriptive label to help identify its meaning. Additionally, as we'll soon see, we can also label individual values of a variable. Here's how we add the variable label:

🖰   Move the cursor into the **Label** column, and type Name of Swimmer. As you type, notice that the column gets wider. This completes the definition of our first variable.

🖰   Now let's create a variable to represent gender. Move to the first column of row 2, and name the new variable Gender.

🖰   Although gender is qualitative, we are going to represent it numerically. Since Numeric is the default in the **Type** column, we'll skip that column and go to **Width**. We are using a one-digit number to represent gender, but for ease of reading, let's make the width 4 by clicking on the downward pointing arrow (see below)[2].

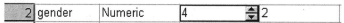

🖰   Our coded values are integers; in the **Decimal** column, reduce the 2 to 0.

🖰   **Label** this variable Sex of swimmer.

🖰   Now we can assign text labels to our coded values. In the **Values** column, click the gray box with three dots. This opens the **Value Labels** dialog box (completed version shown here). Type 1 in the Value box and type Female in the Value Label box. Click **Add**.

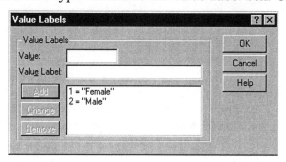

---

[2] Another reason to allow additional space is to avoid a problem with the number of decimal spaces. By default, a numeric variable is eight spaces wide, with two decimal positions. To reduce the width to two spaces, we'd need to first reduce the decimal positions.

🖰 Then type 2 in Value, and Male in Value Label. Click **Add**, and then click **OK**.

Finally, we'll create a *scale* variable in this dataset: Time.

🖰 Begin as you have done twice now, by naming the third variable Time. You may leave **Type** as it is, since the default setting of 8 spaces wide with two decimal places is appropriate here.[3]

🖰 **Label** this variable "Practice time (secs)."

🖰 Switch to the Data View by clicking the appropriate tab in the lower left of your screen.

Follow the directions below, using the data table found on page 3. If you make a mistake, just return to the cell and retype the entry.

🖰 Move the cursor to the first cell below Swimmer, and type Sara; then press Enter. In the next cell, and type Jason. When you've completed the names, move to the top cell under Gender, and go on. When you are done, the Data Editor should look like this:

	swimmer	gender	time	var
1	Sara	1	29.34	
2	Jason	2	30.98	
3	Joanna	1	29.78	
4	Donna	1	34.16	
5	Phil	2	39.66	
6	Hanna	1	44.38	
7	Sam	2	34.80	
8	Ben	2	40.71	
9	Abby	1	37.03	
10	Justin	2	32.81	

🖰 In the **View** menu at the top of your screen, select **Value Labels**; do you see the effect in the Data Editor? Return to the **View** menu and click **Value Labels** again. You can toggle labels on and off in this way.

---

[3] When we create a numeric variable, we specify the maximum length of the variable and the number of decimal places. For example, the data type "Numeric 8.2" refers to a number eight characters long, of which the final two places follow the decimal point: e.g., 12345.78.

Saving a Data File

It is wise to save all of your work in a disk file. SPSS distinguishes between two types of files—output and data—that one might want to save. At this point, we've created a data file and ought to save it on a disk. Let's call the data file **Swim.**

> 🖳 Check with your instructor to see if you can save the data file on a hard drive or network drive in your system.

> 🖱 On the **File** menu, choose **Save As....** In the **Save in** box, select **3½ Floppy (A:)**. Then, next to **File Name**, type swim. Click **Save.**

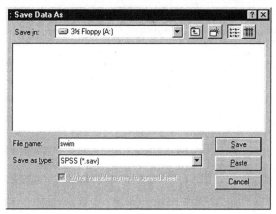

Creating a Bar Chart

With the data entered and saved, we can begin to look for an answer for the coach. We'll first use a bar graph to display the average time for the males in comparison to the females.

> 🖱 Click on **Graphs** in the menu bar, and choose **Bar....** You will see the dialog box shown to the right.

As you can see, there are several options available. This is true for many commands; we'll typically use the default options early in this book, moving to other choices as you become more familiar with statistics and with SPSS.

🖑 Click on the button marked **Define**.

Now you should see this dialog box; complete it following the numbered steps, which are explained further below.

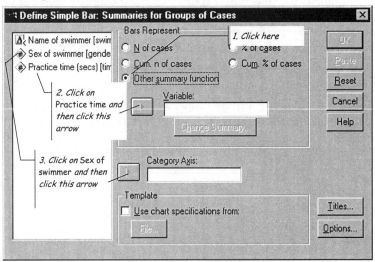

Notice that the three variables are initially listed by description and name on the left side of the dialog box, along with special symbols:

    A̪   Nominal variable (qualitative)

    ⊕   Scale variable (quantitative)

🖑 Click the circle marked Other summary function.

🖑 In the list of variables, click on Practice time [time], and then click the arrow button next to the box marked Variable. Notice that the expression Mean(Practice time (secs) [time]) appears in the box; the chart will display the mean, or average, practice times.

🖑 Now click on Sex of swimmer, and click the arrow next to the box marked Category Axis. Then click **OK**.

You will now see a new window appear, containing a bar chart (see facing page). This is the output Viewer, and contains two "panes." On the left is the Outline pane, which displays an outline of all of your output. The Content pane, on the right, contains the output itself.

Also, notice the menu bar at the top of the Viewer window. It is very similar to the one in the Data Editor, with some minor differences. In general, we can perform statistical analysis from either window. Later, we'll learn some data manipulation commands that can only be given from the Data Editor.

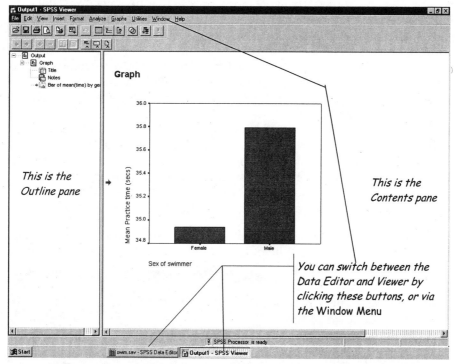

The first thing you should do is to put your name on your output. Eventually when you print the output, you will be able to identify yours.

> From now on in this book, we'll abbreviate menu selections with the name of the menu and the submenu or command. The command you just gave would be **Graphs ➤ Bar...**

Insert ➤ New Page Title On the Output pane, a box will appear. Type your name in it, and then move the cursor just outside of the box.[4] Click the left mouse button.

Before evaluating the bar chart, look at the bottom of your screen. There are two buttons with SPSS icons on them. One says swim.sav - SPSS Data Editor, and the other Output1 - SPSS Viewer. You can switch between the Data Editor and the Output Viewer by clicking on one button or the other.

Now look at the chart. The height of each bar corresponds to the simple average time of the males and females. **What does the chart tell you about the original question: Did the males or females have a better practice that day?**

There is more to a set of data than its average. Let's look at another graph that can give us a feel for how the swimmers did individually and collectively. This graph is called a box-and-whiskers plot (or boxplot), and displays how the swimmers' times were spread out. Boxplots are fully discussed in Session 4, but we'll take a first look now. You may issue this command either from the Data Editor or the Viewer.

Graphs ➤ Boxplot... As with the bar chart, you get a small dialog box asking for the type of boxplot we want. We'll use the default, and just click Define.

Complete this dialog box as shown to the right, and click OK.

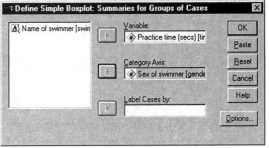

This command generates two outputs, as shown on the facing page. The boxplot is part of the SPSS Explore procedure, so this output is titled **Explore**. The Case Processing Summary is found in many SPSS commands. It lets us know how many cases in the sample have valid values for our variable (Time). Here, we have no missing data.

The Case Processing Summary is followed by the boxplot, which shows results for the males and females. There are two boxes, and each

---

[4] If you click near the graph itself, you will select the graph and a box will appear around it. Later you'll learn about selecting parts of your output. To deselect the graph, click anywhere outside of the selected box.

has "whiskers" extending above and below the box. In this case, the whiskers extended from the shortest to the longest time. The outline of the box reflects the middle three times, and the line through the middle of the box represents the *median* value for the swimmers.[5]

Explore
Sex of swimmer

Case Processing Summary

		Cases					
		Valid		Missing		Total	
	Sex of swimmer	N	Percent	N	Percent	N	Percent
Practice time (secs)	Female	5	100.0%	0	.0%	5	100.0%
	Male	5	100.0%	0	.0%	5	100.0%

Practice time

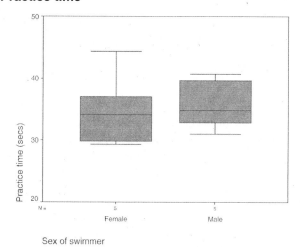

Sex of swimmer

Looking now at the boxplot, what impression do you have of the practice times for the male and female swimmers? How does this compare to your impression from the first graph?

Saving an Output File

At this point, we have the Viewer open with some output and the Data Editor with a data file. We have saved the data, but have not yet saved the output on a disk.

---

[5] The median of a set of points is the middle value when the observations are ranked from smallest to largest.

🖱 **File ➤ Save As...** In this dialog box, assign a name to the file (such as Session 1). This new file will save both the Outline and Content panes of the Viewer.

Getting Help

You may have noticed the **Help** button in the dialog boxes. SPSS features an extensive on-line Help system. If you aren't sure what a term in the dialog box means, or how to interpret the results of a command, click on Help. You can also search for help on a variety of topics via the Help menu at the top of your screen. As you work your way through the sessions in this book, Help may often be valuable. Spend some time experimenting with it before you genuinely need it.

Printing in SPSS

Now that you have created some graphs, let's print them. Be sure that no part of the outline is highlighted; if it is, click once in a clear area of the Outline pane. If a portion of the outline is selected, only that portion will print.

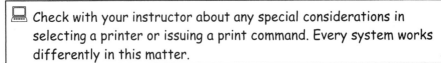 Check with your instructor about any special considerations in selecting a printer or issuing a print command. Every system works differently in this matter.

🖱 **File ➤ Print...** This command will print the Contents pane of the Viewer. Click **OK**.

Quitting SPSS

When you have completed your work, it is important to exit the program properly. Virtually all Windows programs follow the same method of quitting.

🖱 **File ➤ Exit** You will generally see a message asking if you wish to save changes. Since we saved everything earlier, click **No**.

That's all there is to it. Later sessions will explain menus and commands in greater detail. This session is intended as a first look; you will return to these commands and others at a later time.

Session 2

Tables and Graphs for One Variable

Objectives

In this session, you will learn to do the following:
- Retrieve data stored in a SPSS data file
- Explore data with a Stem-and-Leaf display
- Create and customize a histogram
- Create a frequency distribution
- Print output from the Viewer window
- Create a bar chart

Opening a Data File

In the previous session, you created a SPSS data file by entering data into the Data Editor. In this lab, you'll use several data files that are available on your disk. This session begins with some data about traffic accidents in the United States. Our goal is to get a sense of how prevalent fatal accidents were in 1994.

> 🖳 NOTE: The location of SPSS files depends on the configuration of your computer system. Check with your instructor.

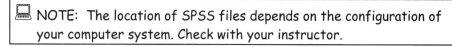 Choose **File ➤ Open ➤ Data...** A dialog box like the one shown on the next page will open. In the Look in: box, select the appropriate directory for your system or network, and you will see a list of available worksheet files. Select the one named **States**. (This file

name may appear as States.sav on your screen, but it's the same file.)

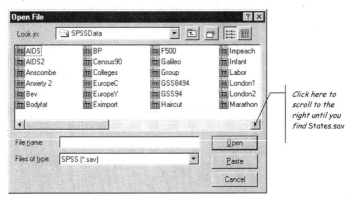

Click here to scroll to the right until you find States.sav

Click **Open**, and the Data Editor will show the data from the **States** file. Using the *scroll bars* at the bottom and right side of the screen, move around the worksheet, just to look at the data. Move the cursor to the row containing variable names (e.g. state, pay93, pay94, etc.) Notice that the *variable labels* appear as the cursor passes each variable name. Consult Appendix A for a full description of the data files.

Exploring the Data

SPSS offers several tools for exploring data, all found in the Explore command. To start, we'll use the Stem-and-Leaf plot to look at the number of people killed in automobile accidents in 1994.

🖰 **Analyze ➤ Descriptive Statistics ➤ Explore...** Select Auto Accident Fatalities [accfat] as shown in this dialog box. Click **OK**.

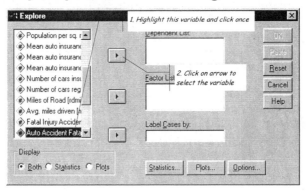

Many SPSS dialog boxes show a list of variables, as this one does. Here the variables are listed in the same order as in the Data Editor. In other dialog boxes, they may be listed alphabetically by *variable label*. When you move your cursor into the list, the entire label becomes visible. The *variable name* appears in square brackets after the label. This book often refers to variables by name, rather than by label. If you cannot find the variable you are looking for, consult Appendix A.

By default, the Explore command reports on the extent of missing data, generates a table of descriptive statistics, creates a stem-and-leaf plot, and constructs a box-and-whiskers plot. The descriptive statistics and boxplot are treated later in Session 4.

Explore

Case Processing Summary

	Cases					
	Valid		Missing		Total	
	N	Percent	N	Percent	N	Percent
Auto Accident Fatalities	51	100.0%	0	.0%	51	100.0%

The first item in the Viewer window summarizes how many observations we have in the dataset; here there are 51 "cases," or observations, in all. For every one of the 50 states plus the District of Columbia, we have a valid data value, and there is no missing data.

Below that is a table of descriptive statistics. For now, we bypass these figures, and look at the Stem-and-Leaf plot, shown here and explained below.

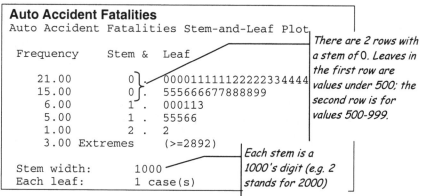

Auto Accident Fatalities
Auto Accident Fatalities Stem-and-Leaf Plot

```
   Frequency     Stem &   Leaf

      21.00          0 .   000011111122222334444
      15.00          0 .   555666677888899
       6.00          1 .   000113
       5.00          1 .   55566
       1.00          2 .   2
       3.00 Extremes     (>=2892)

 Stem width:        1000
 Each leaf:         1 case(s)
```

There are 2 rows with a stem of 0. Leaves in the first row are values under 500; the second row is for values 500-999.

Each stem is a 1000's digit (e.g. 2 stands for 2000)

In this output, there are three columns of information, representing frequency, stems, and leaves. Looking at the notes at the bottom of the plot, we find that each stem line represents a 1000's digit, and each leaf represents 1 state. Let's take a close look at the first row of output to understand what it means.

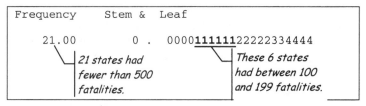

Note that the first two rows both have a 0 stem. The first row represents states with between 0 and 499 fatalities, and the second row represents states with 500 to 999 fatalities. Thus, in the first row of output we find that 21 states had between 0 and 499 automobile accident fatalities in 1994. There are four "0-leaves" in that first row; these represent four states that had fewer than 100 fatalities that year. The six "1-leaves" (highlighted above) represent six states with between 100 and 199 fatalities. Moving down the plot, the row with a stem of 2 indicates that one state had between 2000 and 2499 fatalities. Finally, in the last row, we find that 3 states had at least 2892 fatalities, and that these are considered extreme values.

The Stem-and-Leaf plot helps us to represent a body of data in a comprehensible way, and permits us to get a feel for the "shape" of the distribution. It can help us to develop a meaningful frequency distribution, and provides a crude visual display of the data. For a better visual display, we turn to a histogram.

Creating a Histogram

In the first session, we created some simple graphs. In this session, we'll begin to learn about SPSS's interactive graphs.

⌐ **Graphs ➤ Interactive ➤ Histogram....** As shown on the facing page, select accfat by clicking on it and dragging it to the horizontal axis (see next page). Notice that in this dialog box, the variables are listed alphabetically by *variable label*, unlike the Explore dialog box.

⌐ Click on the Titles tab, and type a title for this graph (e.g. "1994 Traffic Fatalities"), place your name in the caption, and click **OK**. Your histogram will appear in the Viewer window.

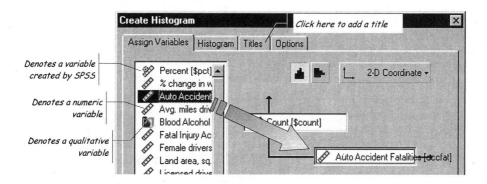

Denotes a variable created by SPSS

Denotes a numeric variable

Denotes a qualitative variable

1994 Traffic Fatalities

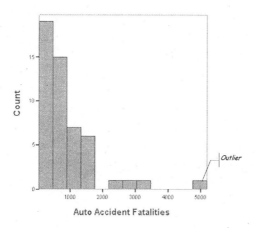

Prepared by R. Carver & J. Nash

The horizontal axis represents a number of fatalities, and the vertical represents the number of states reporting that many cases. The histogram provides a visual sense of the frequency distribution. Notice that the vast majority of the states appear on the left end of the graph. *How would you describe the shape of this distribution?* Compare this histogram to the Stem-and-Leaf plot. *What important differences, if any, do you see?*

Also notice the very short bar at the extreme right end of the graph. *What state do you think it might represent?* Look in the Data Editor to find that outlier.

In this histogram, SPSS determined the number of bars, which affects the apparent shape of the distribution. In an interactive histogram we can control the number of bars as follows:

🖰 Double-click anywhere on your histogram. There are several buttons displayed across the top and down the left side of the graph. Click on the Chart Manager button, as shown here.

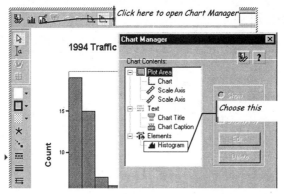

🖰 When the Chart Manager dialog box appears, highlight **Histogram** in the outline, and then click **Edit**. You will then see the Histogram dialog box shown here.

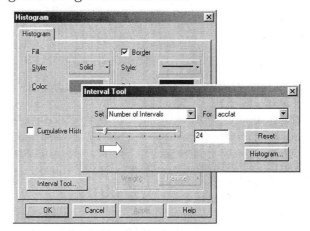

🖰 Click on the **Interval Tool** box, and use the slider control to specify 24 intervals. You may need a steady hand to stop the slider precisely at 24; be patient and careful.

🖱 Click **Histogram**, then click **OK**.

🖱 Close the Chart Manager dialog box by clicking on the ⊠ button in the upper right corner.

How does this compare to your first histogram? Which graph better summarizes the dataset? Explain.
We would expect more populous states to have more fatalities than smaller states. As such, it might make more sense to think in terms of the *proportion* of the population killed in accidents in each state. In our dataset, we have a variable called Traffic fatalities per 100,000 pop [ratefat].

🖱 Construct a histogram for the variable Ratefat. Note that you can replace Accfat with Ratefat by dragging the new variable into the horizontal axis position within the Histogram dialog box.

🖱 Click on the title tab, and notice that the title of the previous graph is still there. Replace it with a new title.

Looking at the histogram, how would you describe the shape of this distribution? What was the approximate average rate of fatalities per 100,000 residents in 1994? Is there an outlier as in the earlier analysis? In which states are traffic fatalities most prevalent?

🖱 Now, return to the interactive Histogram dialog box. At the lower left-center of the dialog box, find and check Cumulative Histogram; click **OK**.

A cumulative histogram displays cumulative frequency. As you read along the horizontal axis from left to right, the height of the bars represents the number of states experiencing a rate less than or equal to the value on the horizontal axis. Compare the results of this graph to the prior graph. **About how many states had traffic fatality rates of less than 20 fatalities per 100,000 population?**

Frequency Distributions

Let's look at some questions concerning *qualitative data.* Switch from the Viewer window back to the Data Editor window.

🖱 **File ➤ Open ➤ Data...** Choose the data file called **Census90.** You'll notice that the **States** file automatically closes when you open a different file.

This file contains a random sample of 982 Massachusetts citizens, with their responses to selected questions on the 1990 United States Decennial Census. One question on the census form asked how they commute to work. In our dataset, the relevant variable is called Means of transportation [trans]. This is a *categorical*, or *nominal*, variable. The Bureau of Census has assigned the following code numbers to represent the various categories:

Value	Meaning
0	n/a, not a worker or in the labor force
1	Car, Truck, or Van
2	Bus or trolley bus
3	Streetcar or trolley car
4	Subway or elevated
5	Railroad
6	Ferryboat
7	Taxicab
8	Motorcycle
9	Bicycle
10	Walked
11	Worked at Home
12	Other Method

To see how many people in the sample used each method, we can have SPSS generate a simple frequency distribution.

🖱 **Analyze ➤Descriptive Statistics ➤ Frequencies...** Select the variable Means of transportation [trans] and click **OK**.

In the Viewer window, you should now see the output shown on the next page.

Frequencies

Statistics

Means of transportation to work

N	Valid	982
	Missing	0

Means of transportation to work

		Frequency	Percent	Valid Percent	Cumulative Percent
Valid	n/a, not working	556	56.6	56.6	56.6
	Car, truck, van	354	36.0	36.0	92.7
	Bus or trolley bus	10	1.0	1.0	93.7
	Subway or elevated	1	.1	.1	93.8
	Railroad	2	.2	.2	94.0
	Taxicab	2	.2	.2	94.2
	Bicycle	1	.1	.1	94.3
	Walked	28	2.9	2.9	97.1
	Worked at home	22	2.2	2.2	99.4
	Other	6	.6	.6	100.0
	Total	982	100.0	100.0	

Among people who work, which means of transportation is the most common? The least common? Be careful: the most common response was "not working" at all.

Another Bar Chart

To graph this distribution, we should make a bar chart.

🖱 **Graphs ➤ Interactive ➤ Bar...** This dialog box is very similar to the Histogram dialog box. Drag the Trans variable to the X-axis box, title and place your name on the graph, and click **OK**.

The bar chart and frequency distribution should contain the same information. Do they? Comment on the relative merits of using a frequency table versus a bar chart to display the data.

Printing Session Output

Sometimes you will want to print all or part of a Viewer window. Before printing your session, be sure you have typed your name into the output. To print the entire session, click anywhere in the Contents pane of the Viewer window (be sure not to select a portion of the output), and then choose **File ➤ Print**. To print *part* of a Viewer window, do this:

🖱 In the Outline pane of the Viewer window (the left side of the screen), locate the first item of the output that you want to print. Position the cursor on the name of that item, and click the left mouse button.

🖱 Using the scroll bars (if necessary), move the cursor to the end of the portion you want to print. Then press Shift on the keyboard and click the left mouse button. You'll see your selection highlighted, as shown here.

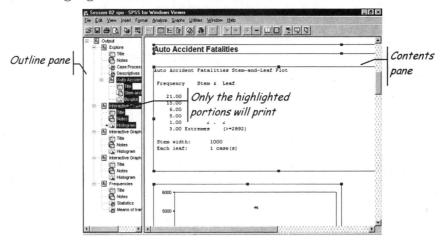

🖱 **File ➤ Print...** Notice that the Selection button is already marked, meaning that you'll print a selection of the output within the Contents pane. Click **OK**.

Moving On...

Using the skills you have practiced in this session, now answer the following questions. In each case, provide an appropriate graph or table to justify your answer, and *explain* how you drew your conclusion.

1. (**Census90** file) Note that the Trans variable includes the responses of people who don't have jobs. Among those who do have jobs, what proportion use some type of public transportation (bus, subway, or railroad)?

For the following questions, you will need to use the files **States**, **Marathon**, **AIDS**, **BP**, and **Nielsen** (see Appendix A for detailed file descriptions). You may be able to use several approaches or commands to answer the question; think about which approach seems best to you.

States

2. The variable named bac refers to the legal blood alcohol threshold for driving while intoxicated. All states set the threshold at either .08 or .10. About what percentage of states use the .08 standard?

3. The variable called ins94 is the average auto insurance premium for each state in 1994. Do drivers in all states pay about the same amount for insurance? What seems to be a typical amount? How much variation is there across states?

4. The variable called mileage is the average number of miles driven per year by a state's drivers. With the help of a Stem-and-Leaf plot, locate (in the Data Editor) two states where drivers lead the nation in miles driven; what are they?

Marathon

This file contains the finish times for the wheelchair racers in the 100th Boston Marathon.

5. The variable Country is a three-letter abbreviation for the home country of the racer. Not surprisingly, most racers were from the USA. What country had the second highest number of racers?

6. Use a cumulative histogram to determine approximately what percentage of wheelchair racers completed the 26-mile course in less than 2 hours, 10 minutes (130 minutes).

7. How would you characterize the shape of the histogram of the variable Minutes? (Experiment with different numbers of intervals in this graph.)

AIDS

This file contains data related to the incidence of AIDS around the world.

8. How would you characterize the shape of the distribution of AIDS cases reported in 1994? Are there any outlying countries? If so, what are they?

9. Consider the 1993 rate of cases reported per 100,000 people. Compare the shape of this distribution to the shape of the distribution in the previous question.

BP

This file contains data about blood pressure and other vital signs for subjects after various physical and mental activities.

10. The variable sbprest is the subject's systolic blood pressure at rest. How would you describe the shape of the distribution of systolic blood pressure for these subjects?

11. Using a cumulative histogram, approximately what percent of subjects had systolic pressure of less than 140?

12. The variable dbprest is the subject's diastolic blood pressure at rest. How would you describe the shape of the distribution of diastolic blood pressure for these subjects?

13. Using a cumulative histogram, approximately what percent of subjects had diastolic pressure of less than 80?

Nielsen

This file contains the Nielsen ratings for the 20 most heavily watched television programs for the week ending September 14, 1997.

14. Which of the networks reported had the most programs in the top 20? Which had the fewest?

15. Approximately what percentage of the programs enjoyed ratings in excess of 11.5?

Tables and Graphs for Two Variables

Objectives

In this session, you will learn to do the following:

- Cross-tabulate two variables
- Create several bar charts comparing two variables
- Create a histogram for two variables
- Create an XY scatterplot for two quantitative variables

Cross-Tabulating Data

The prior session dealt with displays of a single variable. This session covers some techniques for creating displays that compare two variables. Our first example considers two qualitative variables. The example involves the Census data that you saw in the last session, and in particular addresses the question: "Do men and women use the same methods to get to work?" Since sex and means of transportation are both categorical data, our first approach will be a *joint frequency table*, also known as a *cross-tabulation*.

    Open the Census file by selecting **File ➤ Open ➤ Data...**, and choosing **Census90**.

    Analyze ➤ Descriptive Statistics ➤ Crosstabs... In the dialog box (next page), select the variables Means of transportation to work [trans] and Sex [sex], and click **OK**. You'll find the cross-tabulation in the Viewer window. ***Who makes greater use of***

cars, trucks, or vans: Men or women? Explain your reasoning.

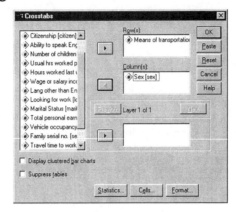

Crosstabs

Case Processing Summary

	Cases					
	Valid		Missing		Total	
	N	Percent	N	Percent	N	Percent
Means of transportation to work * Sex	982	100.0%	0	.0%	982	100.0%

Means of transportation to work * Sex Crosstabulation

Count

		Sex		Total
		Male	Female	
Means of transportation to work	n/a, not working	258	298	556
	Car, truck, van	186	168	354
	Bus or trolley bus	3	7	10
	Subway or elevated	1		1
	Railroad	2		2
	Taxicab		2	2
	Bicycle		1	1
	Walked	15	13	28
	Worked at home	6	16	22
	Other	3	3	6
Total		474	508	982

The results of the **Crosstabs** command are not mysterious. The case processing summary indicates that there were 982 cases, with no missing data. In the crosstab itself, the rows of the table represent the various means of transportation, and the columns refer to males and females. Thus, for instance, 168 women commuted in a car, truck, or van.

Simply looking at the frequencies could be misleading, since the sample does not have equal numbers of men and women. It might be more helpful to compare the percentage of men commuting in this way to the percentage of women doing so. Even percentages can be misleading if the samples are small. Here, fortunately, we have a large sample. Later we'll learn to evaluate sample information more critically with an eye toward sample size.

The cross-tabulation function can easily convert the frequencies to relative frequencies. We could do this by returning to the Crosstabs dialog box following the same menus as before, or by taking a slightly different path.

Editing a Recent Dialog

Often, we'll want to repeat a command using different variables or options. For quick access to a recent command, SPSS provides a special button on the toolbar below the menus. Click on the **Dialog Recall** button (shown to the right), and you'll see a list of recently issued commands. **Crosstabs** will be at the top of the list; click on Crosstabs, and the last dialog box will reappear.

To answer the question posed above, we want the values in each cell to reflect frequencies relative to the number of women and men, so we want to divide each by the total of each respective column. To do so, click on the button marked **Cells**, check Column Percentages, click **Continue**, and then click **OK**. *Based on this table, would you say that men or women are more likely to commute by car, truck, or van?*

Now try asking for Row Percentages (click on **Dialog Recall**). *What do these numbers represent?*

More on Bar Charts

We can also use a bar chart to analyze the relationship between two variables. Let's look at the relationship between two *qualitative* variables in the student survey: gender and seat belt usage. Students were asked how frequently they wear seat belts when driving: Never, Sometimes, Usually, and Always. What do you think the students said? Do you think males and females responded similarly? We will create a bar chart to help answer these questions.

🖰 In the Data Editor, open the file called **Student**.

🖰 **Graphs ➤ Interactive ➤ Bar...** We used this command in the prior session. We must specify a variable for the horizontal axis, and may optionally specify other variables.

🖰 Drag Frequency of seat belt usage [belt] to the horizontal axis box. If we were to click **OK** now, we would see the total number of students who gave each response. But we are interested in the comparison of responses by men and women.

🖰 In the Legend Variables area, specify that you want bar colors assigned according to Gender. Now click **OK**.

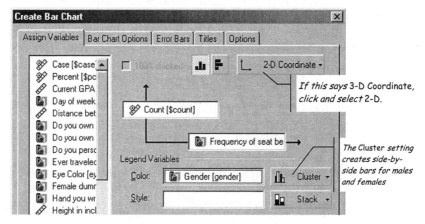

The resulting graph appears at the bottom of the next page. ***What can you say about the seat belt habits of these students?***

In this bar chart, the order of axis categories is alphabetical. With this variable, it would be more logical to have the categories sequenced by frequency: Never, Sometimes, Usually, and Always.

🖰 As you learned in Session 2, click on the interactive bar chart, and press the **Chart Manager** button. In the Chart Manager, select the Categorical Axis, and click **Edit** (see next page).

🖰 Place the cursor on the word Always and drag it to the end of the Exclude Categories list. Notice that, under Order in the dialog box, the radio button labeled By custom order is selected. Click

OK, and close the Chart Manager. The resulting graph should make a bit more sense.

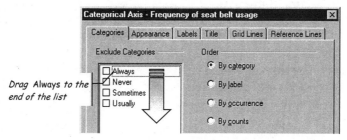

This graph uses clustered bars to compare the responses of the men and the women. A clustered bar graph highlights the differences in belt use by men and women, but it's hard to tell how many students are in each usage category. A stacked bar chart is a useful alternative.

✲ **Graphs ➤ Interactive ➤ Bar...** Return to the interactive Bar Chart dialog box (shown on the previous page).

✲ In the Legend Variables area, click on the button now marked Cluster, and select Stack instead.

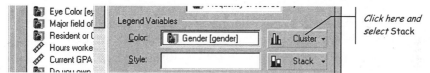

Here are the clustered and stacked versions of this graph. ***Do they show different information? What impressions would a viewer draw from these graphs?***

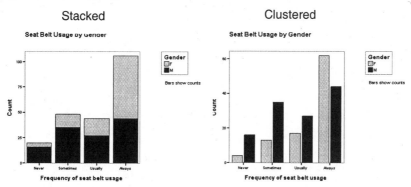

We can also analyze a quantitative variable in a bar chart. Let's compare the grade point averages (GPA) of the men and women in the student survey. We might compare the averages of the two groups.

🖱 **Graphs ➤ Interactive ➤ Bar...**

🖱 Drag Count [$count] *from* the vertical axis back to the list of available variables on the left side of the dialog box. Drag Current GPA [gpa] to the vertical axis and Gender to the horizontal axis.

Look at the lower portion of the dialog box; it now looks like this:

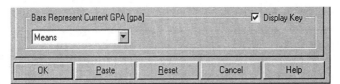

The text box indicates the bars in the graph will represent the mean, or average, of the GPA variable. Click **OK**, and look at the bar chart. ***How do the average GPAs of males and females compare?***

Comparing Two Distributions

The bar chart compared the mean GPAs for men and women. How do the whole distributions compare? As a review, we begin by looking at the distribution of GPAs for all students.

🖱 **Graphs ➤ Interactive ➤ Histogram...** Select Current GPA [gpa] as the variable, and click **OK**. You'll see the graph shown here. ***How do you describe the shape of this distribution?***

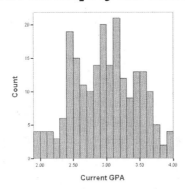

Let's compare the distribution of grades for male and female students. We'll create two side-by-side histograms, using the same vertical and horizontal scales:

🖱 Click the **Dialog Recall** button, and choose Create Histogram. We need to indicate that the graph should distinguish between the GPAs for women and men.

🖱 Drag the variable Gender into Panel Variables area, and then click **OK** in the dialog box.

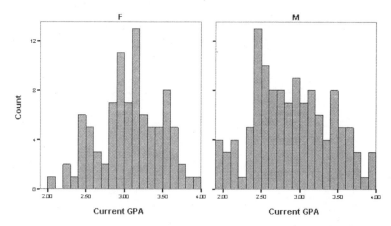

What does this graph show about the GPAs of these students? In what ways are they different? What do they have in common? What reasons might explain the patterns you see?

Scatterplots to Detect Relationships

The prior example involved a quantitative and a qualitative variable. Sometimes, we might suspect a connection between two quantitative variables. In the student data, for example, we might think that taller students generally weigh more than shorter ones. We can create a scatterplot or XY graph to investigate.

🖱 **Graphs ➤ Interactive ➤ Scatterplot...** In the dialog box (next page), select Weight in pounds [wt] as the *y*, or vertical axis variable, and Height in inches [ht] as the *x* variable. Click **OK**.

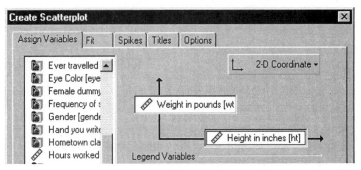

Look at the scatterplot, reproduced here. **Describe what you see in the scatterplot. By eye, approximate the range of weights of students who are 5'2" (or 62 inches) tall. Roughly how much more do 6'2" students weigh?**

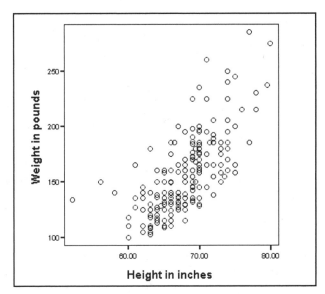

🖰 We can easily incorporate a third variable into this graph. Return to the scatterplot dialog box, and drag Gender to the box marked Color in the Legend Variables area. Click **OK.**

In what ways is this graph different from the first scatterplot? What additional information does it convey? What generalizations can you make about the heights and weights of men and women? Which points might we consider to be outliers?

Moving On...

Create the tables and graphs described below. Refer to Appendix A for complete data descriptions. ***Be sure to title each graph, including your name. Print the completed graphs.***

Student

1. Generate side-by-side histograms of the distribution of heights, separating men and women. Comment on the similarities and differences between the two groups.

2. Do the same for students' weights.

Bev

3. Using the **Interactive Bar Chart** command, display the mean of Revenue per Employee, by SIC category. Which beverage industry generates the highest average revenue by employee?

4. Make a similar comparison of Inventory Turnover averages. How might you explain the pattern you see?

SlavDiet

In *Time on the Cross: The Economics of American Negro Slavery,* by Robert William Fogel and Stanley Engerman, the diets of slaves and the general population are compared.

5. Create two bar charts summing up the calories consumed by each group, by food type. How did the diets of slaves compare to the rest of the population, according to these data? [NOTE: you want the bars to represent the *sum* of calories. After dragging the calories variable to the vertical axis, look at the Bars Represent area of the dialog box. It will say Means. Click on the drop-down list, and choose Sums.]

Galileo

In the 16th century, Galileo conducted a series of famous experiments concerning gravity and projectiles. In one experiment, he released a ball to roll down a ramp. He then measured the total horizontal distance which the ball traveled until it came to a stop. The data from that experiment occupy the first two columns of the data file.

In a second experiment, a horizontal shelf was added to the base of the ramp, so that the ball rolled directly onto the shelf from the ramp. Galileo recorded the vertical height and horizontal travel for this apparatus as well, which are in the third and fourth column of the file.[1]

6. Construct a scatterplot for the first experiment, with release height on the x axis and horizontal distance on the y axis. Describe the relationship between x and y.

7. Do the same for the second experiment.

AIDS

8. Construct a bar chart that displays the mean AIDS case rate in 1993, by World Health Organization region. Which region of the world had the highest incidence of AIDS cases per 100,000 population in 1993?

Mendel

Gregor Mendel's early work laid the foundations for modern genetics. In one series of experiments with several generations of pea plants, his theory predicted the relative frequency of four possible combinations of color and texture of peas.

9. Construct bar charts of both the actual experimental (observed) results and the predicted frequencies for the peas. Comment on the similarities and differences between what Mendel's theory predicted, and what his experiments showed.

Salem

In 1692, twenty persons were executed in connection with the famous witchcraft trials in Salem, Massachusetts. At the center of the controversy was Rev. Samuel Parris, minister of the parish at Salem Village. The teenage girls who began the cycle of accusations often gathered at his home, and he spoke out against witchcraft. This data file

---

[1] *Sources:* Drake, Stillman. *Galileo at Work,* (Chicago: University of Chicago Press, 1978); Dickey, David A. and Arnold, J. Tim "Teaching Statistics with Data of Historic Significance," *Journal of Statistics Education,* v.3, no. 1, 1995.

represents a list of all residents who paid taxes to the parish in 1692. In 1695, many villagers signed a petition supporting Rev. Parris.

10. Construct a crosstab of proParris status and the accuser variable. (Hint: Compute row or column percents, using the **Cells** button.) Based on the crosstab, is there any indication that accusers were more or less likely than nonaccusers to support Rev. Parris? Explain.

11. Construct a crosstab of proParris status and the defend variable. Based on the crosstab, is there any indication that defenders were more or less likely than nondefenders to support Rev. Parris? Explain.

12. Create a chart showing the mean (average) taxes paid, by accused status. Did one group tend to pay higher taxes than the other? If so, which group paid more?

Impeach

This file contains the results of the U.S. Senate votes in the impeachment trial of President Clinton in 1999.

13. The variable called conserv is a rating scale indicating how conservative a senator is (0 = very liberal, 100 = very conservative). Use a bar chart to compare the mean ratings of those who cast 0, 1, or 2 votes to convict the President. Comment on any pattern you see.

14. The variable called Clint96 indicates the percentage of the popular vote cast for President Clinton in the senator's home state in the 1996 election. Use a bar chart to compare the mean percentages for those senators who cast 0, 1, or 2 votes to convict the President. Comment on any pattern you see.

GSS94

These questions were selected from the 1994 General Social Survey. For each, construct a crosstab and discuss any possible relationship indicated by your analysis.

15. Does a person's political outlook (liberal vs. conservative) appear to vary by their highest educational degree?

16. One question asks if a bad marriage is better than no marriage at all. Did women and men tend to respond similarly? Did married and unmarried people tend to respond similarly?

17. One question asks respondents about how frequently they have sex. Did men and women respond similarly?

18. How does attendance at religious services vary by region of the country?

GSS8494

This file contains responses to a series of General Social Survey questions from 1984 and 1994. Respondents were different in the two years. Use an interactive bar chart to display the percentages of responses to the following questions, comparing the 1984 and 1994 results. Comment on the changes, if any, you see in the ten-year comparison.

19. Should marijuana be legalized?

20. Should abortion be allowed if a woman wants one for any reason?

21. Should colleges permit racists to teach?

22. Are you afraid to walk in your neighborhood at night?

States

23. Use a scatterplot to explore the relationship between the number of fatal accidents in a state and the population of the state. Comment on the pattern, if any, in the scatterplot.

24. Use a scatterplot to explore the relationship between the number of fatal accidents in a state and the mileage driven within the state. Comment on the pattern, if any, in the scatterplot.

Nielsen

25. Chart the mean (average) rating by network. Comment on how well each network did that week. (Refer to your work in Session 2.)

Session 4

One-Variable Descriptive Statistics

Objectives

In this session, you will learn to do the following:
- Compute measures of central tendency and dispersion for a variable
- Create a box-and-whiskers plot for a single variable
- Compute z-scores for all values of a variable

Computing One Summary Measure for a Variable

There are several measures of central tendency (mean, median, and mode) and of dispersion (range, variance, standard deviation, etc.) for a single variable. You can use SPSS to compute these measures. We'll start with the mode of an *ordinal* variable.

  Open the data file called **Student**. The variables in this file are student responses to a first-day-of-class survey.

One variable in the file is called Drive. This variable represents students' responses to the question, "How would you rate yourself as a driver?" The answer codes are as follows:

> 1 = Below average
> 2 = Average
> 3 = Above Average

We'll begin by creating a frequency distribution:

🖑 **Analyze ➤ Descriptive Statistics ➤ Frequencies...** Scroll down the list of variables until you find How do you rate your driving? [drive]. Select the variable, and click **OK**. Look at the results. *What was the modal response? What strikes you about this frequency distribution? How many students are in the "middle"? Is there anything peculiar about these students' view of "average"?*

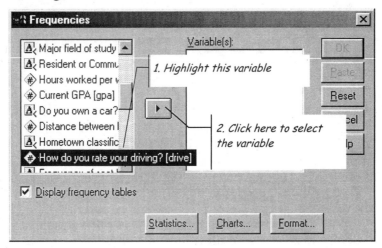

Frequencies

Statistics

How do you rate your driving?

N	Valid	218
	Missing	1

One student did not answer

How do you rate your driving?

		Frequency	Percent	Valid Percent	Cumulative Percent
Valid	Below Average	8	3.7	3.7	3.7
	Average	106	48.4	48.6	52.3
	Above Aveage	104	47.5	47.7	100.0
	Total	218	99.5	100.0	
Missing	System	1	.5		
Total		219	100.0		

What does each column above tell you?

Drive is a qualitative variable with three possible values. Some categorical variables have only two values, and are known as *binary* variables. Gender, for instance, is binary. In this dataset, there are two variables representing a student's sex. The first, which you have seen in earlier sessions, is called Gender, and assumes values of F and M. The second is called Female dummy variable [female], and is a numeric variable equal to 0 for men and 1 for women. We call such a variable a *dummy variable* since it artificially uses a number to represent categories. If we wanted to know the *proportion* of women in the sample, we could tally either variable. Alternatively, we could compute the mean of Female. By summing all of the 0s and 1s, we would find the number of women; dividing by *n* would yield the sample proportion.

🖰 **Analyze ➤ Descriptive Statistics ➤ Descriptives...** In this dialog box, scroll down and select Female dummy variable [female], and click **OK**.

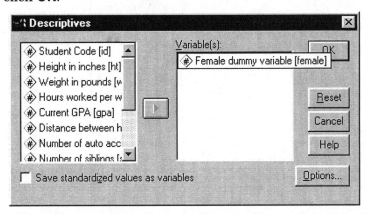

According to the Descriptives output, 44% of these students were females. Now let's move on to a *quantitative* variable: the number of brothers and sisters the student has. The variable is called sibling.

🖰 **Analyze ➤ Descriptive Statistics ➤ Frequencies...** Select the variable Number of siblings [sibling]. Click on **Statistics**, and select Quartiles, Mean, Median, and Mode. Click **Continue**, then **OK**.

Requesting these options generates the output shown on the next page. You probably are familiar with mean, median, and mode. *Quartiles* divide the data into four equal groups. Twenty-five percent of the observations fall below the first quartile, and 25% fall above the third quartile. The second quartile is the same as the median.

Frequencies

Statistics

Number of siblings

N	Valid	218
	Missing	1
Mean		1.80
Median		1.00
Mode		1
Percentiles	25	1.00
	50	1.00
	75	2.00

218 of 219 students answered this question.

Number of siblings

		Frequency	Percent	Valid Percent	Cumulative Percent
Valid	0	21	9.6	9.6	9.6
	1	90	41.1	41.3	50.9
	2	58	26.5	26.6	77.5
	3	29	13.2	13.3	90.8
	4	11	5.0	5.0	95.9
	5	6	2.7	2.8	98.6
	6	1	.5	.5	99.1
	8	1	.5	.5	99.5
	11	1	.5	.5	100.0
	Total	218	99.5	100.0	
Missing	System	1	.5		
Total		219	100.0		

Compare the mean, median, and mode. As summaries of the "center" of the dataset, what are the relative merits of these three measures? If you had to summarize the answers of the 218 students, which of the three would be most appropriate? Explain.

We can also compute the mean with the **Descriptives** command, which provides some additional information about the dispersion of the data.

⌐∅ **Analyze ➤ Descriptive Statistics ➤ Descriptives...** As you did earlier for female, find the mean for sibling.

Now look in the Viewer window, and you will see the mean number of siblings per student. Note that you now see the sample standard deviation, the minimum, and the maximum as well.

Computing Additional Summary Measures

By default, the **Descriptives** command provides the sample size, minimum, maximum, mean, and standard deviation. What if you were interested in another summary descriptive statistic? You could click on the **Options** button within the Descriptives dialog box, and find several other available statistics (try it now).

Alternatively, you might use the **Explore** command to generate a variety of descriptive measures of your data. To illustrate, we'll explore the heights and weights of these students.

Analyze ➤ Descriptive Statistics ➤ Explore... Select the variables Height in inches [ht] and Weight in pounts [wt], as shown in the dialog box below. These will be the dependent variables for now[1]. The **Explore** command can compute descriptive statistics and also generate graphs, which we will see shortly. For now, let's confine our attention to statistics; select **Statistics** in the Display portion of the dialog box, and click **OK**.

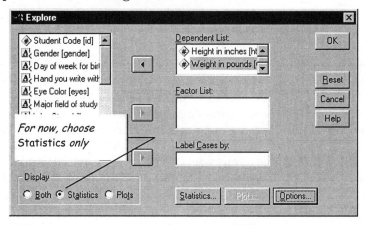

---

[1] When we begin to analyze relationships between two variables, the distinction between dependent variables and factors will become important to us. For the time being, the Dependent List is merely the list of variables we want to describe or explore.

Below is part of the output you'll see (we have omitted the full descriptive information for weight). The output provides a variety of different descriptive statistics for each of the two variables.

Case Processing Summary

	Cases					
	Valid		Missing		Total	
	N	Percent	N	Percent	N	Percent
Height in inches	215	98.2%	4	1.8%	219	100.0%
Weight in pounds	215	98.2%	4	1.8%	219	100.0%

Descriptives

		Statistic	Std. Error
Height in inches	Mean	68.0872	.2834
	95% Confidence Interval for Mean — Lower Bound	67.5286	
	95% Confidence Interval for Mean — Upper Bound	68.6457	
	5% Trimmed Mean	68.0846	
	Median	68.0000	
	Variance	17.508	
	Std. Deviation	4.1843	
	Minimum	52.00	
	Maximum	80.00	
	Range	28.00	
	Interquartile Range	5.0000	
	Skewness	-.067	.165
	Kurtosis	.811	.328

Once again, we start with a Case Processing Summary. Here we find the size of the sample and information about missing cases if any. Four students did not provide information either about their height, their weight, or both. This is an example of accounting for missing data "listwise." For four students, this list of two variables is incomplete. Actually, there was one student who did not report height, and three who did not report weight. If we wanted to compare the data about height and weight, we would have to omit all four students, since they didn't provide complete information.

The table of descriptives shows statistics and standard errors for several statistics. You'll study standard errors later in your course. At this point, let's focus on the statistics. Specifically, SPSS computes these summary measures:

Mean	The sample mean, or $\bar{x} = \dfrac{\sum x}{n}$
95% Confidence Interval for Mean (Lower and Upper Bound)	A confidence interval is a range used to estimate a population mean. You will learn how to determine the bounds of a confidence interval later in the course.
5% Trimmed Mean	The 5% trimmed sample mean, computed by omitting the highest and lowest 5% of the sample data.[2]
Median	The sample median (50th percentile)
Variance	The sample variance, or $s^2 = \dfrac{\sum(x-\bar{x})^2}{n-1}$
Std. Deviation	The sample standard deviation, or the positive square root of s^2.
Minimum	The minimum observed value for the variable
Maximum	The maximum observed value for the variable
Range	Maximum–minimum
Interquartile range	The third quartile (Q3, or 75th percentile) minus the first quartile (Q1, or 25th percentile) for the variable.
Skewness	Skewness is a measure of the symmetry of a distribution. A perfectly symmetrical distribution has a skewness of 0, though a value of 0 does not necessarily indicate symmetry. If the distribution is skewed to the right or left, skewness is positive or negative, respectively.
Kurtosis	Kurtosis is a measure of the general shape of the distribution, and can be used to compare the distribution to a normal distribution later in the course.

In your Viewer window, compare the mean, median, and trimmed mean for the two variables. ***Does either of the two appear to have some outliers skewing the distribution?*** Reconcile your conclusion with the skewness statistic.

The Explore command offers several graphical options which relate the summary statistics to the graphs you worked with in earlier labs. For example, let's take a closer look at the heights.

🖰 Return to the Explore dialog box, and select **Plots** in the Display area. Click **OK**.

---

[2] If a faculty member computes your grade after dropping your highest and lowest scores, she is computing a trimmed mean.

By default, this will generate a stem-and-leaf display and a box-and-whiskers plot. Look at the stem-and-leaf display for heights: ***Does it confirm your judgement about the presence or absence of outliers?***

A Box-and-Whiskers Plot

The **Explore** command generates the *five-number summary* for a variable (minimum, maximum, first and third quartiles, and median). A boxplot, or box-and-whiskers plot displays the five number summary.[3] Additionally, it permits easy comparisons, as we will see. The box in the center of the plot shows the interquartile range, with the median located by the dark horizontal line. The "whiskers" are the t-shaped lines extending above and below the box; nearly all of the students lie within the region bounded by the whiskers. The few very tall and short students are identified individually by labeled circles.

A boxplot of a single variable is not terribly informative. Here is an alternative way to generate a boxplot, this time creating two side-by-side graphs for the male and female students.

🖰 **Graphs ➤ Boxplot...** In the Boxplot dialog box (see below left), choose Simple and click **Define**. In the next dialog box, select Height in inches as the variable, and Gender for the Category Axis. Click **OK**.

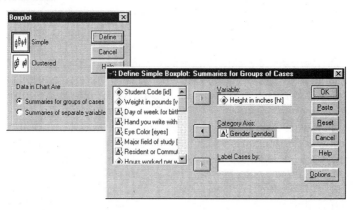

---

[3] Actually, the whiskers in a SPSS boxplot may not extend to the minimum and maximum values. The lines project from the box at most a length of 1.5 times the interquartile range (IQR). Outliers are represented by labeled circles, and extreme values (more than 3 times the IQR), by asterisks.

How do the two resulting boxplots compare to one another? What does the graph suggest about the center and spread of the height variable for these two groups? Comment on what you see.

🖱 Let's try another boxplot; return to the Define Simple Boxplot dialog box.

🖱 Click on Height, and notice that the arrow button now points to the left. Click the arrow to remove Height as the variable and replace it with Weight, but retain Gender as the category axis variable.

How do the two boxplots for weight compare? How do the weight and height boxplots compare to one another? Can you account for the differences?

Standardizing a Variable

🖱 Now open the file called **Marathon**. This file contains the finish times for all wheelchair racers in the 100th Boston Marathon.

🖱 Find the mean and median finish times. *What do these two statistics suggest about the symmetry of the data?*

Since many of us don't know much about wheelchair racing or marathons, it may be difficult to know if a particular finish time is good or not. It is sometimes useful to *standardize* a variable, so as to express each value as a number of standard deviations above or below the mean. Such values are also known as *z-scores*.

🖱 **Analyze ➤ Descriptive Statistics ➤ Descriptives...** Select the variable Finish times [minutes]. Before clicking **OK**, check the box marked Save standardized values as variables. This will create a new variable representing each racer's z-score.

Now look at the Data Editor; notice a new variable, zminutes. Since the racers are listed by finish rank, the first z-score value belongs to the winner of the race, whose finishing time was well below average. That's why his z-score is negative, indicating that his time was less than the mean.

Locate the racer with a z-score of approximately 0. **What does that z-score indicate about this racer?**

Look at the z-scores of the top two racers. **How does the difference between them compare to the difference between finishers #2 and #3? Between the last two finishers?**

Statisticians think of ratio variables (such as Minutes or zminutes) as containing more information than ordinal variables (such as Rank). **How does this example illustrate that difference?**

Moving On...

Now use the commands illustrated in this session to answer these questions. Where appropriate, indicate which statistics you computed, and why you chose to rely on them to draw a conclusion.

Student

1. What was the mean amount paid for a haircut?

2. What was the median amount paid for a haircut?

3. Comment on the comparison of the mean and median.

Colleges

This file contains tuition and other data from a 1994 survey of colleges and universities in the United States.

4. In 1994, what was the average in-state tuition [Tuit_in] at U.S. colleges? Out-of-state tuition? [Tuit_out]. Is it better to look at means or medians of these particular variables? Why?

5. Which varies more: in-state or out-of-state tuition? Why is that so? (Hint: Think about how you should measure variation.)

6. Standardize the in-state tuition variable. Find your school in the Data Editor (schools are listed alphabetically within state). What is the z-score for your school, and what does the z-score tell you?

Output

This file contains data concerning industrial production in the United States from 1945–1996. Capacity utilization, all industries represents the degree to which the productive capacity of all U.S. industries was utilized. Capacity utilization, mfg has a comparable figure, just for manufacturers.

7. During the period in question, what was the mean utilization rate for all industrial production? What was the median? Characterize the symmetry and shape of the distribution for this variable.

8. During the period in question, what was the mean utilization rate for manufacturing? What was the median? Describe the symmetry and shape of the distribution for this variable.

9. In terms of their standard deviations, which varied more: overall utilization or manufacturing utilization?

10. Comment on similarities and differences between the center, shape, and spread of these two variables.

Sleep

This file contains data about the sleeping patterns of different animal species.

11. Construct box-and-whiskers plots for Lifespan and Sleep. For each plot, explain what the landmarks on the plot tell you about each variable.

12. The mean and median for the Sleep variable are nearly the same (approximately 10.5 hours). How do the mean and median of Lifespan compare to each other? What accounts for the comparison?

13. According to the dataset, "Man" (row 34) has a maximum life span of 100 years, and sleeps 8 hours per day. Refer to a boxplot to approximate, in terms of quartiles, where humans fall among the species for each of the two variables.

14. Sleep hours are divided into two types: dreaming and nondreaming sleep. On average, do species spend more hours in dreaming sleep or nondreaming sleep?

Water

These data concern water usage in 221 regional water districts in the United States for 1985 and 1990.

15. The 17th variable, Total freshwater consumptive use 1985 [tocufr85], is the total amount of fresh water used for consumption (drinking) in 1985. On average, how much drinking water did regions consume in 1985?

16. One of the last variables, Consumptive use % of total use 1985 [pctcu85] is the percentage of all fresh water devoted to consumptive use (as opposed to irrigation, etc.) in 1985. What percentage of fresh water was consumed, on average, in water regions during 1985?

17. Which of the two distributions was more heavily skewed? Why was that variable less symmetric than the other?

BP

These data include blood pressure measurements from a sample of students after various physical and psychological stresses.

18. Compute measures of central tendency and dispersion for the resting diastolic blood pressure. Do the same for diastolic blood pressure following a mental arithmetic activity. Comment on the comparison of central tendency, dispersion, and symmetry of these two distributions.

19. Compute measures of central tendency and dispersion for the resting systolic blood pressure. Do the same for systolic blood pressure following a mental arithmetic activity. Comment on the comparison of central tendency, dispersion, and symmetry of these two distributions.

Two-Variable Descriptive Statistics

Objectives

In this session, you will learn to do the following:

- Compute the coefficient of variation
- Compute measures of central tendency and dispersion for two variables or two groups
- Compute the covariance and correlation coefficient for two quantitative variables

Comparing Dispersion with the Coefficient of Variation

In the previous session, you learned to compute descriptive measures for a variable, and to compare these measures for different variables. Often, the more interesting statistical questions require us to compare two sets of data, or to explore possible relationships between two variables. This session introduces techniques for making such comparisons and describing such relationships.

Comparing the means or medians of two variables is straightforward. On the other hand, when we compare the dispersion of two variables, it is sometimes helpful to take into account the magnitude of the individual data values. For instance, suppose we sampled the heights of mature maple trees and corn stalks. We could anticipate the standard deviation for the trees to be larger than that of the stalks, simply because the heights themselves are so much larger. In general, variables with large means may tend to have large dispersion. What we need is a *relative measure* of dispersion. That is what the coefficient of variation (CV) is. The CV is the standard deviation expressed as a percentage of the mean. Algebraically, it is:

$$CV = 100 \cdot \left(\frac{s}{\overline{x}} \right)$$

Unfortunately, SPSS does not have a command to compute the coefficient of variation for a variable in a data file. Our approach will be to have SPSS find the mean and standard deviation, and simply compute the CV by hand.

🖰 Open the file called **Coldges**.

> 💻 Beginning with this session, we will begin to drop the instruction "Click **OK**" at the end of each dialog. Only when there is a sequence of commands in a dialog box will you see Click **OK**.

🖰 **Analyze ➤ Descriptive Statistics ➤ Descriptives...** Select the variables In-state tuition (tuit_in) and Out-of-state tuition (tuit_out). These values are different for state colleges and universities, but for private schools they are usually the same. Not surprisingly, the mean for out-of-state tuition exceeds that for in-state.

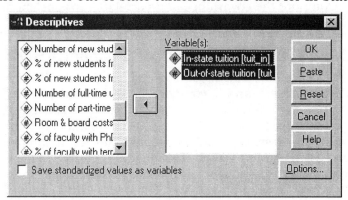

But notice the standard deviations. **Which variable varies more? Why is that so?** The comparison is all the more interesting when we look at the coefficient of variation.

Using your hand calculator, find the coefficient of variation for both variables. **What do you notice about the degree of variation for these two variables? Does in-state tuition vary a little more or a** *lot more than out-of-state? What real-world reasons could account for the differences in variation?*

Descriptive Measures for Subsamples

Residency status is just one factor in determining tuition. Another important consideration is the difference between public and private institutions. We have a variable called PubPvt which equals 1 for public (state) schools, and 2 for private schools. In other words, the PubPvt column represents a qualitative attribute of the schools. We can compute separate descriptive measures for these two groups of institutions. To do so, we invoke the **Explore** command:

🖱 **Analyze ➤ Descriptive Statistics ➤ Explore...** As shown in the dialog box, select the two tuition variables as the Dependent List, and Public/Private School as the Factor List. In the Display area, select Statistics, and click **OK**.[1]

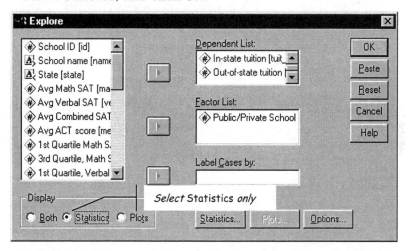

Look in the Viewer window for the numerical results. These should look somewhat familiar, with one new twist. For each variable, two sets of output appear. The first refers to those sample observations with a PubPvt value of 1 (i.e., the State schools); the second refers to the Private school subsample. Take a moment to familiarize yourself with the

---

[1] You can achieve similar results with the **Analyze ➤ Compare Means ➤ Means...** command. As is often the case, there are many ways to approach our data in SPSS.

output. Compute the CVs for each of the four sets of output; *relatively speaking, where is dispersion the greatest?* Find your school in the dataset. *For in-state tuition, is your school above or below the mean among comparable schools?*

Measures of Association: Covariance and Correlation

We have just described a relationship between a quantitative variable (Tuition) and a qualitative variable (Public vs. Private). Sometimes, we may be interested in a possible relationship or association between two *quantitative* variables. For instance, in this dataset, we might expect that there is a relationship between the number of admissions applications a school receives (AppsRec) and the number of new students it accepts for admission (AppsAcc).

🖰 **Graphs ➤ Scatter...** Choose a simple scatterplot, click on **Define**, and place AppsAcc on the *Y* axis, and AppsRec on the *x* axis. *Do you see evidence of a relationship? How would you describe it?*

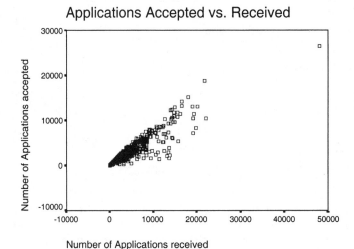

Applications Accepted vs. Received

A graph like this one shows a strong tendency for *x* and *y* to covary. In this instance, schools with higher *x* values also tend to have higher *y* values. Given the meaning of the variables, this makes sense.

There are two common statistical measures of covariation. They are the *covariance* and the *coefficient of correlation*. In both cases, they are computed using all available observations for a pair of variables. The formula for the sample covariance of two variables, x and y, is this:

$$\text{cov}_{xy} = \frac{\sum (x_i - \overline{x})(y_i - \overline{y})}{n-1}$$

The sample correlation coefficient[2] is:

$$r = \frac{\text{cov}_{xy}}{s_x s_y}$$

where:

s_x, s_y are the sample standard deviations of x and y, respectively.

In general, we confine our interest to correlation, computed as follows:

🖰 **Analyze ➤ Correlate ➤ Bivariate...** Select the variables AppsRec and AppsAcc, and click **OK**. You will see the results in your Viewer window (next page).

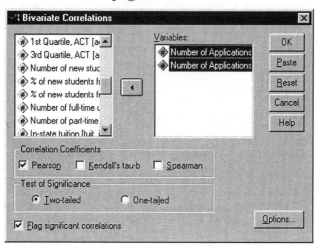

_____

[2] Formally, this is the Pearson Product Moment Correlation Coefficient, known by the symbol, r.

Correlations

		Number of Applications received	Number of Applications accepted
Number of Applications received	Pearson Correlation	1.000	.931**
	Sig. (2-tailed)	.	.000
	N	1292	1289
Number of Applications accepted	Pearson Correlation	.931**	1.000
	Sig. (2-tailed)	.000	.
	N	1289	1291

**. Correlation is significant at the 0.01 level (2-tailed).

The two equal values highlighted in this image are the sample correlation between applications received and applications accepted, based on 1289 schools. The notation Sig. (2-tailed) will be explained in Session 11; at this point, it is sufficient to say that a Sig. value of .000 indicates a statistically meaningful correlation.

By definition, a correlation coefficient (symbol r) assumes a value between -1 and +1. Absolute values near 1 are considered strong correlations; that is, the two variables have a strong tendency to vary together. This table shows a strong correlation between the variables. Absolute values near 0 are weak correlations, indicating very little relationship or association between the two variables.

Variables can have strong sample correlations for many possible reasons. It may be that one causes the other (or vice versa), that a third variable causes both of them, or that their observed association in this particular sample is merely a coincidence. As you will learn later in your course, correlation is an important tool in statistical reasoning, but we must never assume that correlation implies causation.

Moving On...

Use the commands and techniques presented in this session to answer the following questions. Explain your choice of statistics in responding to each question.

Impeach

This file contains data about the U.S. senators who voted in the impeachment trial of President Clinton.

1. Compare the mean of the percentage vote for Clinton in the 1996 election for Republican and Democratic senators, and comment on what you find.

2. What is the correlation between the number of votes a senator cast against the President in the trial and the number of years left in the senator's term? Comment on the strength of the correlation.

GSS94

These are data extracted from the 1994 General Social Survey.

3. Did female respondents tend to watch more or less television per day than male respondents?

4. One question on the survey asks if the respondent is afraid to walk alone in the neighborhood. Compare the mean ages of those who said "yes" to those who said "no."

World90

This file contains economic and population data from 42 countries around the world. These questions focus on the distribution of Gross Domestic Product (GDP) in the countries.

5. Compare the means of C, I, and G (the proportion of GDP committed to consumption, investment, and government, respectively). Which is highest, on average? Why might that be?

6. Compare the mean and median for G. Why do they differ so?

7. Compare the coefficients of variation for C and for I. Which varies more: C or I? Why?

8. Compute the correlation coefficient for C and I. What does it tell you?

F500

This worksheet contains data about the 1996 Fortune 500 companies.

9. How strong an association exists between profit and revenue among these companies? (Hint: Find the correlation.)

10. One of the variables is MktVal, representing the market value of the firm as of March 15, 1996. Presumably, several financial factors are related to market value. Which of the four seems to have the strongest relationship to market value: Revenue, Profits, Equity, or Growth (1985–95)? Explain your rationale, referring to statistical evidence.

Bev

This is the worksheet with data about the beverage industry.

11. If you have studied accounting, you may be familiar with the current ratio, and what it can indicate about the firm. What is the mean current ratio in this sample of beverage industry firms? (See Appendix A for a definition of current ratio.)

12. In the entire sample, is there a relationship between the current and quick ratios? Why might there be one?

13. How do the descriptive measures for the current and quick ratios compare across the SIC subgroups? Suggest some possible reasons for the differences you observe.

Bodyfat

This dataset contains body measurements of 252 males.

14. What is the sample correlation coefficient between neck and chest circumference? Suggest some reasons underlying the strength of this correlation.

15. What is the sample correlation coefficient between biceps and forearm? Suggest some reasons underlying the strength of this correlation.

16. Which of the following variables is most closely related to bodyfat percentage (FatPerc): age, weight, abdomen circumference, or thigh circumference? Why might this be?

Salem

These are the data from Salem Village, Massachusetts in 1692. Refer to Session 3 for further description. Using appropriate descriptive and graphical techniques, compare the average taxes paid in the three groups listed below. In each case, explain whether you should compare means or medians, and state your conclusion.

17. Defenders vs. nondefenders

18. Accusers vs. nonaccusers

19. Rev. Parris supporters vs. nonsupporters

Sleep

This worksheet contains data about the sleep patterns of various mammal species. Refer back to Session 4 for more information.

20. Using appropriate descriptive and graphical techniques, how would you characterize the relationship (if any) between the amount of sleep a species requires and the mean weight of the species?

21. Using appropriate descriptive and graphical techniques, how would you characterize the relationship (if any) between the amount of sleep a species requires and the life span of the species?

Water

In Session 4, you computed descriptive measure for the Total freshwater consumptive use 1985 (tocufr85). The 33rd variable (tocufr90) contains comparable data for 1990.

22. Compare the means and medians for these columns. Did regions consume more or less water, on average, in 1990 than they did in 1985? What might explain the differences five years later?

23. Compare the coefficient of variation for each of the two variables. In which year were the regions more varied in their consumption patterns? Why might this be?

24. Construct a scatterplot of freshwater consumptive use in 1990 versus the regional populations in that year. Also, compute the correlation coefficient for the two variables. Is there evidence of a relationship between the two? Explain your conclusions, and suggest reasons for the extent of the relationship (if any).

Elementary Probability

Objectives

In this session, you will learn to do the following:
- Simulate random sampling from a population
- Draw a random sample from a set of observations
- Manipulate worksheet data for analysis

Simulation

Thus far, all of our work with SPSS has relied on observed sets of data. Sometimes we will want to exploit the program's ability to *simulate* data which conforms to our own specifications. In the case of experiments in classical probability, for instance, we can have SPSS simulate flipping a coin 10,000 times, or rolling a die 500 times.

A Classical Example

Imagine a game spinner with four equal quadrants, such as the one illustrated here. Suppose you were to record the results of 1000 spins. **What do you expect the results to be?**

We can simulate 1000 spins of the spinner by having SPSS calculate some pseudorandom data:

🖱 **File ➤ Open ➤ Data...** Retrieve the data file called **Spinner**. This file has two variables: The first, spin, is simply a list running from 1 to 1000. The second, quadrant, has 1000 missing values.

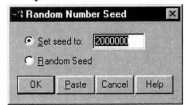

🖱 **Transform ➤ Random Number Seed...** Whenever SPSS generates pseudorandom values, it uses an algorithm that requires a "seed" value to begin the computations. By default, SPSS uses 2,000,000 as the seed value; you should type in your own seed value, choosing any whole number between 1 and 2 billion. When you do, the dialog box will vanish with no visible effect; the consequences of this command become apparent shortly.

🖱 **Transform ➤ Compute...** Complete the dialog box exactly as shown below. The command uses two functions, RV.UNIFORM and TRUNC. RV.UNIFORM(1,5) will randomly generate real numbers greater than 1 and less than 5. TRUNC truncates the number, leaving the integer portion. This gives us random integers between 1 and 4, simulating our spinner.

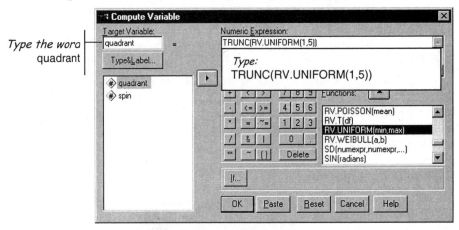

As soon as you click **OK**, you will see this message:

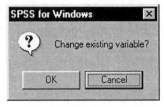

This message cautions you that you are about to replace the missing values in quadrant with new values; you should click **OK**, creating a random sample of 1,000 spins. The first column identifies the trial spin, and the second contains random values from 1 to 4.

> 🖥 NOTE: Because these are random data, your data will be unique. If you are working with a partner on another computer, her results will differ from yours.

🖰 **Analyze ➤ Descriptive Statistics ➤ Frequencies...** Create a frequency distribution for the variable named quadrant. What should the relative frequency be for each value? ***Do all of your results exactly match the theoretical value? To the extent that they differ, why do they?***

Recall that classical probabilities give us the long-run relative frequency of a value. Clearly, 1000 spins is not the "long-run," but this simulation may help you understand what it means to say that the probability of spinning any single value equals 0.25.

Observed Relative Frequency as Probability

As you know, many random events are not classical probability experiments, and we must rely on observed relative frequency. In this part of the session, we will direct our attention to the Census data, and focus on the chance that a randomly selected individual speaks a language other than English at home. The Census asked "Do you speak a language other than English at home?" These respondents gave three different answers: 0 indicates the individual did not answer or was under 5 years old; 1 indicates that the respondent spoke another language; and 2 indicates that the respondent spoke only English at home.

🖰 Open the **Census90** data file. When you give this command, you'll be asked if you want to save the changes in **Spinner**; say "no."

🖰 **Analyze ➤ Descriptive Statistics ➤ Frequencies...** Choose the variable Lang other than English (lang1).***What do these relative frequencies (i.e. percents) indicate?***

Lang other than English

		Frequency	Percent	Valid Percent	Cumulative Percent
Valid	n/a, under 5 years	66	6.7	6.7	6.7
	Yes, speaks other language	64	6.5	6.5	13.2
	No, English only	852	86.8	86.8	100.0
	Total	982	100.0	100.0	

If you were to choose one person from the 982 who answered this question, what is the probability that the person does speak a language other than English at home? Which answer are you most likely to get?

Suppose we think of these 982 people as a population. What kind of results might we find if we were randomly to choose 50 people from the population, and tabulate their answers to the question? Would we get exactly 86.8% speaking English only?

In SPSS, we can randomly select a sample from a data file. To access the relevant command, we must be in the Data Editor. Switch to the Data Editor now.

Data ➤ Select Cases... We want to sample 50 rows from the dataset, and then look at the frequencies for Lang1. Complete the dialog box as shown here:

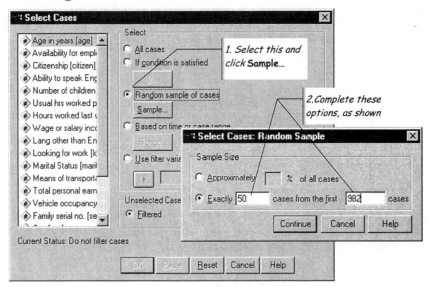

Look in the Data Editor. Note that many case numbers are crossed out, indicating that nearly all of the cases were not selected. Also, if you scroll to the right-most column of the dataset, you will find a new variable (filter_$) that equals 0 for excluded cases, and 1 for included cases. Let's now find the results for the 50 randomly chosen cases.

🖱 **Analyze ➤ Descriptive Statistics ➤ Frequencies...** In the Viewer window, there is a frequency distribution for your random sample of 50 individuals. *How did these people respond? How similar is this distribution to the entire population of 982 people?*

Before drawing the random sample, we know that almost 87% of all respondents speak only English. Knowledge of the relative frequency is of little value in predicting the response of one person, but it is quite useful in predicting the overall results of asking 50 people.

Handling Alphanumeric Data

In the prior example, the variable of interest was numeric. What if the variable is not represented numerically in the dataset?

🖱 Open the file called **Colleges**. Imagine choosing one of these colleges at random. *What's the chance of choosing a college from California?*

We could create a frequency table of the state names, and find out how many schools are in each state. That will give us a very long frequency table. Instead, let's see how to get a frequency table that just classifies all schools as being in California or elsewhere. To do so, we can first create a new variable, differentiating California from nonCalifornia schools. This requires several steps. First, switch to the Data Editor, and proceed as follows:

🖱 **Transform ➤ Recode ➤ Into Different Variables...** We will create a new variable (Calif), coded as Calif for California colleges, and Other for colleges in all other states (see dialog boxes, next page).

🖱 From the variable list, select State.

🖱 In the Output Variable area, type Calif in the Name box, California schools in the Label box, and click **Change**.

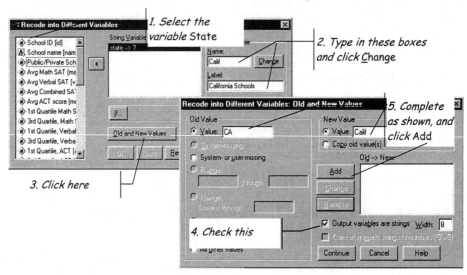

🖱 Click on **Old and New Values...** bringing up another dialog box.

🖱 Complete the dialog boxes as shown above, clicking **Add** to complete that part of the recoding process. After you click **Add**, you'll notice 'CA' → 'Calif' in the Old → New box.

🖱 Now click All other values, and recode them to Other.

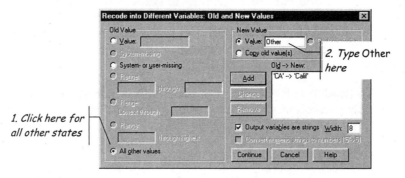

🖰 Click **Add**, then **Continue**. Finally, in the main dialog box, click **OK**.

If you look in the Data Editor window, you'll find the new variable Calif. The first several rows in the column say Other; scroll down to the California schools to see the effect of this command. This is precisely what we want. Now we have a *binary* variable: It equals Calif for schools in California, and Other for schools in all other states. Now we can conveniently figure frequencies and relative frequencies. Similarly, we have the variable Public/Private School [pubpvt] that equals 1 for public or state colleges and 2 for private schools.

Suppose we intend to choose a school at random. In the language of elementary probability, let's define two events. If the randomly chosen school is in California, event C has occurred. If the randomly chosen school is Private, event Pv has occurred. We can cross-tabulate the data to analyze the probabilities of these two events.

🖰 **Analyze ➤ Descriptive Statistics ➤ Crosstabs...** For the rows, select Public/Private School and for the columns, choose California Schools. In the Crosstabs dialog box, click on **Cells**, and check Total Percentages.

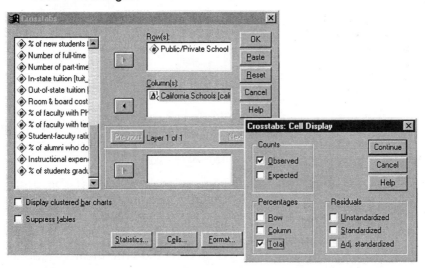

Look at the table in the Viewer window, reproduced on the next page. In the table, locate the twenty-eight California public colleges and universities. **Does California have an unusual proportion of public colleges?**

Public/Private School * California Schools Crosstabulation

			California Schools		Total
			Calif	Other	Total
Public/Private School	Public	Count	28	442	470
		% of Total	2.2%	33.9%	36.1%
	Private	Count	42	790	832
		% of Total	3.2%	60.7%	63.9%
Total		Count	70	1232	1302
		% of Total	5.4%	94.6%	100.0%

Moving On...

Within your current data file and recalling the events just defined, use the **Crosstabs** command and the results shown above to find and comment on the following probabilities:

1. $P(C) = ?$

2. $P(Pv) = ?$

3. $P(C \cap Pv) = ?$

4. $P(C \cup Pv) = ?$

5. $P(Pv \mid C) = ?$

Spinner

Open the **Spinner** data file again, and generate random data as shown earlier, this time with a minimum value of 0, and a maximum value of 2 (this will generate a column of 0s and 1s).

6. What should the mean value of the random data be, and why? Compute and comment on the mean for quadrant.

7. Now have SPSS randomly select 10 cases from the 1000 rows, and compute the mean. Comment on how these results compare to your prior results. Why do the means compare in this way?

8. Repeat the prior question for samples of 100 and 500 cases. Each time, comment on how these results compare to your prior results.

GSS94

This is the excerpt from the 1994 General Social Survey.

Cross-tabulate the responses to the questions about respondent's sex and "who in your household plans the meals?"

9. What is the probability of randomly selecting someone who said "Usually the woman"?

10. What is the probability of randomly selecting a male who said "Usually the woman"?

11. Given that the respondent answered "Usually the woman," what was the probability that the respondent is a woman?

Now cross-tabulate the responses to the questions about the respondent's marital status and the assertion that a bad marriage is better than none at all.

12. What is the probability of randomly selecting a person who is currently married?

13. What is the probability of selecting a married person who disagrees with the statement?

14. What is the probability of selecting a divorced person who disagrees with the statement?

15. What is the probability that a person agrees with the statement, given that the person has never been married?

Discrete Probability Distributions

Objectives

In this session, you will learn to do the following:
- Work with an observed discrete probability distribution
- Compute binomial probabilities
- Compute Poisson probabilities

An Empirical Discrete Distribution

We already know how to summarize observed data; an *empirical distribution* is an observed relative frequency distribution that we intend to use to approximate the probabilities of a random variable. As an illustration, were we to randomly select a woman and ask how many children she has borne, we could regard that number to be a random variable. We will use the data in the **Census90** file to illustrate.

Open the data file **Census90**. Recall that this contains 1990 Census questionnaire responses from Massachusetts residents.

In this file, we are interested primarily in the variable Number of children ever born (fertil), which is defined as the number of children born to women 15 years of age and older. (See Appendix A for a detailed description.) Unfortunately, our dataset also includes men and preadolescent girls. Therefore, before analyzing the data, we need to specify the subsample of cases to use in the analysis.

To do this, we use the **Select Cases** command to choose females over the age of 14.

🖰 **Data ➤ Select Cases...** Choose the If condition is satisfied radio button, and click on the button marked **If....** As shown below, complete the Select Cases: If dialog box to specify that we want only those cases where age > 14 & sex = 1 (i.e. women, aged 15 and older).

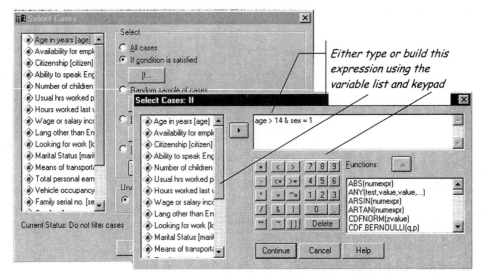

🖰 Click **Continue** in the If dialog box, and then **OK** in the main dialog box.

As in the previous session, this command filters out women under 15 years and all men. Now, any analysis we do will consider only the women 15 and older.

🖰 **Analyze ➤ Descriptive Statistics ➤ Frequencies** Select the variable Number of children ever born (fertil), and click **OK**. Look at the frequency distribution in the Viewer window, directing your attention to the Valid Percent and Cumulative Percent columns.

In terms of probability, what do these percentages mean? If we were to select one woman randomly, what is the probability that we would select a woman who had never given birth? What is the probability that we would select a woman who had given birth to four or fewer children?

Number of children ever born

		Frequency	Percent	Valid Percent	Cumulative Percent
Valid	No child	121	30.0	30.0	30.0
	1 child	71	17.6	17.6	47.6
	2 children	88	21.8	21.8	69.5
	3 children	61	15.1	15.1	84.6
	4 children	38	9.4	9.4	94.0
	5 children	10	2.5	2.5	96.5
	6 children	5	1.2	1.2	97.8
	7 children	4	1.0	1.0	98.8
	8 children	1	.2	.2	99.0
	9 children	2	.5	.5	99.5
	10 children	1	.2	.2	99.8
	12 or more children	1	.2	.2	100.0
	Total	403	100.0	100.0	

Graphing a Distribution

It is often helpful to graph a probability distribution, typically by drawing a line at each possible value of X. The height of the line is proportional to the probability. We'll use the **Bar Chart** command again.

🖑 **Graphs ➤ Interactive ➤ Bar...** In the dialog box, select fertil as the horizontal axis variable, and click **OK**. Then, double-click on the bar chart itself in the Viewer window to activate the interactive commands. Double-click on any bar in the graph to open the Bars dialog box. Select the Bar Width tab, and slide the control all the way to the left (see below). Click **OK**. *Comment on the shape of the distribution.*

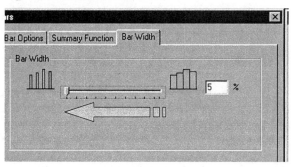

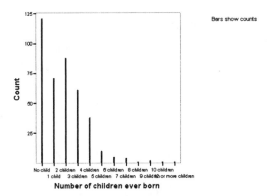

If we were to sample one woman at random, what's the most likely outcome? How many children, on average, did these women report? Which definition of average (mean, median, or mode) is most appropriate here, and why?

A Theoretical Distribution: The Binomial[1]

Some random variables arise out of processes which allow us to specify their distributions without empirical observation. SPSS can help us by either simulating such variables or by computing their distributions. In this lab session, we'll focus on their distributions.

🖰 **File ➤ New ➤ Data...** Create a new data file in the Data Editor.

🖰 Click on the **Variable View** tab and define three new numeric variables (see Session 1 to review this technique). Call the first one x, and define it as numeric 4.0. Call the second variable b25 and the third b40. Specify that each of these is numeric 8.4.

🖰 We will begin by computing the cumulative binomial distribution[2] for an experiment with eight trials and a 0.25 probability of success on each trial. Enter the values 0 through

---

[1] This section assumes you have been studying the binomial distribution in class and are familiar with it. Consult your primary text for the necessary theoretical background.

[2] Some texts provide tables for either Binomial distributions, cumulative Binomial distributions, or both. The cumulative distribution shows $P(X \leq x)$, while the simple distribution shows $P(X = x)$.

8, as shown below, into the first nine cases of x. These values represent the nine possible values of the binomial random variable, *x* or the number of successes in eight trials.

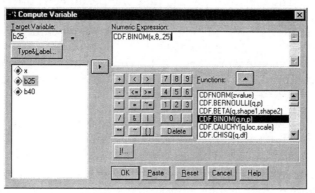

⌐⊕ **Transform ➤ Compute...** In this dialog box, specify that the variable b25 equals the cumulative distribution function for a binomial with eight trials and probability of success of 0.25. Look closely at the Data Editor to see the consequences of this command. Reconcile the data file with the binomial tables in your primary textbook.

⌐⊕ **Graphs ➤ Interactive ➤ Scatterplot...** Your *y* variable is b25 and the *x* variable is x. In the Spikes tab of the dialog box, select Spike to: X1 Axis, and click **OK. *Comment on the shape of this cumulative distribution.***

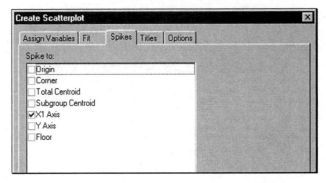

Now we'll repeat this process of a second binomial variable. This time, there are still eight trials, but p(success) = 0.40.

🖰 **Transform ➤ Compute...** Change the target variable to b40, and change the .25 to .40 in the formula.

Look at the Data Editor, and compare b25 to b40. ***Comment on differences. How will the graph of b40 compare to that of b25?***

Another Theoretical Distribution: The Poisson[3]

We can compute several common discrete distributions besides the Binomial distribution. Let's look at one more. The *Poisson* distribution is often a useful model of events which occur over a fixed period of time. The distribution has just one parameter, and that is its mean. In our first binomial example, we had eight trials and a 0.25 probability of success. In the long run, the expected value or mean of x would be 25% of 8, or 2 successes.

Using the same worksheet as for the binomial example, we'll construct the cumulative distribution for a Poisson random variable with a mean of 2 successes. Do the following:

🖰 Create another new variable in the Data Editor. Call it p2, and specify that its type is numeric 8.4.

🖰 **Transform ➤ Compute...** In this dialog box, type in p2 as the Target variable. Replace the Numeric Expression with this:

---

[3] This section assumes you have been studying the Poisson distribution in class and are familiar with it. Consult your primary text for the necessary theoretical background.

CDF.POISSON(x,2), and click **OK**. This expression tells SPSS to compute the cumulative distribution function for a Poisson variable with a mean value of 2, using the number of successes specified in the variable X.

🖱 Plot this variable as you did with the binomial. ***How do these graphs compare to one another?***

Moving On...

Let's use what we have learned to (a) analyze an observed distribution and (b) see how well the Binomial or Poisson distribution serves as a model for the observed relative frequencies.

Student

Students were asked how many automobile accidents they had been involved in during the past two years. The variable called acc records their answers. Perform these steps to answer the question below:

   a) Construct a frequency distribution for the number of accidents.
   b) Find the mean of this variable.
   c) In an empty column of the worksheet, create a variable called X, and type the values 0 through 9 (i.e., 0 in Row 1, 1 in Row 2, etc.).
   d) Create a variable called poisson.
   e) Generate a Poisson distribution with a mean equal to the mean number of accidents. The target variable is poisson, and your numeric expression will refer to X.

   1. Compare the actual cumulative percent of accidents to the Poisson distribution (either visually or graphically). Does the Poisson distribution appear to be a good approximation of the actual data? Comment on the comparison.

Pennies

A professor has his students each flip 10 pennies, and record the number of heads. Each student repeats the experiment 30 times and then records the results in a worksheet.

2. Compare the actual observed results (in a graph or table) with the theoretical Binomial distribution with $n = 10$ trials and $p = 0.5$. Is the Binomial distribution a good model of what actually occurred when the students flipped the pennies? Explain. (*Hint*: Start by finding the mean of each column; since each student conducted 30 experiments, the mean *should* be approximately 30 times the theoretical probability.) NOTE: The actual data will give you an approximation of the simple Binomial probabilities and SPSS will compute the *cumulative* probabilities. When making your comparison, take that important difference into account!

Web

Twenty trials of twenty random queries were made using the Yahoo!® Internet search engine's Random Yahoo! Link. For some links, instead of successfully connecting to a Web site, an error message appeared. In this data file, the variable called problems indicates the number of error messages received in each set of twenty queries. Perform the following steps to answer the questions below:

a) Find the mean of the variable problems and divide it by 20. This will give you a percentage, or probability of success (obtaining an error message in this case) in each query.

b) Create a new variable prob (for number of possible problems encountered) and type the values 0 through 20 (i.e. 0 in Row 1, 1 in Row 2, etc.)

c) Create another new variable called binom, of type Numeric 8.4.

d) Generate a theoretical Binomial distribution with N= 20 (number of trials) and p= probability of success. The target variable is binom and your numeric expression refers to prob.

e) Now produce a cumulative frequency distribution for the variable problems.

3. Compare the actual cumulative percent of problems to the theoretical Binomial distribution. Does the Binomial distribution provide a good approximation of the real data? Comment on both the similarities and differences as well as reasons they might have occurred.

4. Using this theoretical Binomial probability, what is the probability that you will receive exactly three error messages? How many times did this actually occur? Why are there differences? If the sample size was N= 200, what do you think the difference would look like?

Airline

From 1970 to 1997, most major airlines throughout the world have recorded total flight miles their planes have traveled as well as the number of fatal accidents that have occurred. A fatal flight is defined as one in which at least one person (crew or passenger) has died on the flight or as a result of complications on the flight. Perform these steps to answer the questions below:

    a) Create a frequency distribution for the variable events (the number of flights in which a fatality occurred) and find the mean of this variable.

    b) In an empty column, create a new variable x which will represent the number of possible accidents (type 0 through 17, 0 being the lowest observation and 17 being the highest in this sample).

    c) Create a variable called poisson.

    d) Generate a theoretical Poisson distribution with the mean equal to the mean of events. The target variable is poisson and the numeric expression will refer to x.

5. Compare the actual cumulative frequencies to the theoretical cumulative Poisson distribution. Comment on the similarities and differences between the two. Is there anything about the actual observations that surprises you?

6. What do you think the distribution of fatal crashes will look like over the next twenty-seven years? Can the Poisson distribution be used to approximate this observed distribution? What differences between this distribution and one for the future might you expect to see?

Session 8

Probability Density Functions

Objectives

In this session, you will learn to do the following:
- Compute probabilities for any normal random variable
- Use normal curves to approximate other distributions

Continuous Random Variables

The prior session dealt exclusively with *discrete* random variables, that is, variables whose possible values can be listed (such as 0, 1, 2, etc.). In contrast, some random variables are *continuous*. Think about riding in an elevator. As the floor numbers light up on the panel, they do so discretely, in steps as it were: first, then second, and so forth. The elevator, though, is travelling smoothly and continuously through space. We might think of the vertical distance traveled as a continuous variable and floor number as a discrete variable.

The defining property of a continuous random variable is that, for any two values, there are an infinite number of other possible values between them. Between 50 feet and 60 feet above ground level, there are an infinite number of vertical positions the elevator might occupy.

We cannot tabulate a continuous variable as we can a discrete variable, nor can we assign a unique probability to each possible value. This important fact forces us to think about probability in a new way when we are dealing with continuous random variables.

Rather than constructing a probability distribution, as we did for discrete variables, we will use a *probability density function* when dealing with a continuous random variable, x. We'll envision probability as being dispersed over the permissible range of *x*; sometimes the probability is *dense* near particular values, meaning that neighborhood of *x* values is relatively likely. The density function itself is difficult to interpret, but the area beneath the density function[1] represents probability. The area under the entire density function equals 1, and the area between two selected values represents the probability that the random variable falls between those values.

Perhaps the most closely studied family of random variables is the *normal distribution*.[2] We begin this session by considering several specific normal random variables.

Generating Normal Distributions

There are an infinite number of normally distributed random variables, each with its own pair of *parameters*: μ and σ. If we know that *x* is normal with mean μ and standard deviation σ, we know all there is to know about *x*. Throughout this session, we'll denote a normal random variable as $x \sim N(\mu, \sigma)$. For example, $x \sim N(10,2)$ refers to a random variable *x* that is normally distributed with a mean value of 10, and a standard deviation of 2.

The first task in this session will be to specify the density function for three different distributions, to see how the mean and standard deviation define a unique curve. Specifically, we'll generate values of the density function for a *standard normal variable*, $z \sim N(0,1)$, and two others: $x \sim N(1,1)$ and $x \sim N(0,3)$.

   Open the data file called **Normal.** Upon opening the file, you'll see that there is one defined variable (x) that ranges from –8 to +8, increasing with an increment of 0.2. This variable will represent possible values of our random variable.

   Transform ➤ Compute... As shown in the dialog box on the next page, we can compute the cumulative density function for each value of x. Specify that cn01 is the target, and the expression is CDF.NORMAL(x,0,1).

---

[1] For students familiar with calculus, the area under the density function is the integral. You don't need to know calculus or remember the fine points of integration to work with density functions.

[2] As in the prior chapter, we do not provide a full presentation of the normal distribution here. Refer to your primary textbook for more detail.

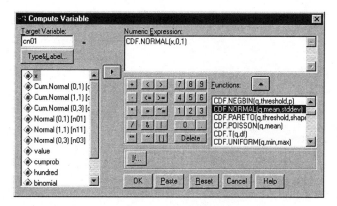

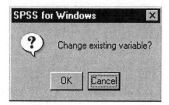

⏻ When you click **OK**, you'll see the message below; click **OK**. Now cn01 contains cumulative density values for $x{\sim}N(0,1)$.

⏻ Now repeat the **Compute** command, changing the target variable to cn11, and the expression to read CDF.NORMAL(x,1,1). This will generate the cumulative density function for $x{\sim}N(1,1)$.

⏻ Return to the Compute dialog, changing the target variable to cn03, and the expression to read CDF.NORMAL(x,0,3). This will generate the cumulative density function for $x{\sim}N(0,3)$.

Now we have three *cumulative density functions*. Later in the exercise, we'll consider these. First, we'll use these results to create the simple normal density functions. To do so, we must transform each variable in the following way[3]:

⏻ **Transform ➤ Compute...** The Target Variable is n01, and the Numeric Expression is cn01-lag(cn01).

---

[3] This function subtracts the cumulative value in adjacent rows. By doing so, we find the height of the density curve at each x value.

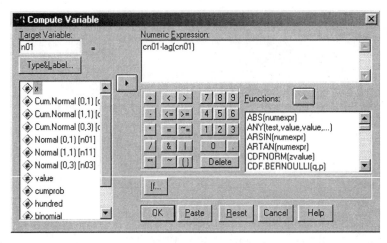

-᷆ Repeat the **Compute** command twice to create n11 and n03, changing the Numeric Expression each time to use cn11 and cn03, respectively.

Now we have the simple density functions for the three normal variables we've been working with. If you were to graph these three normal variables, how would the graphs compare? Where would the curves be located on the number line? Which would be steepest and which would be flattest? Let's see:

-᷆ **Graphs ➤ Line...** In the first dialog box (shown to the right), specify a Multiple line chart, where the lines represent Values of individual cases, as shown. Click **Define**.

-᷆ In the subsequent dialog box (see next page), choose all three normal probability density function variables in the area marked Lines Represent. For Category Labels, click Variable, and choose x. Click **OK**.

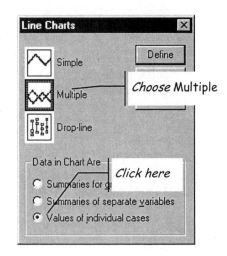

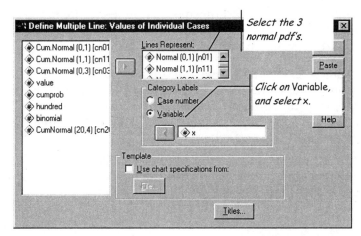

Look at the resulting graph, reproduced here. Your graph will be in color, distinguishing the lines more clearly than this one.

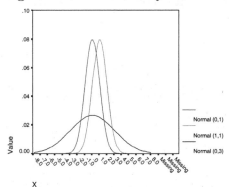

How do these three normal distributions compare to one another? How do the two distributions with a mean of 0 differ? How do the two with a standard deviation of 1 differ?

Finding Areas under a Normal Curve

We often need to compute the probability that a normal variate lies within a given range. We generally convert a variable to the standard normal variable[4], z, consult a table of areas, and then manipulate the areas to find the probability.

---

[4] We make the conversion using the formula $z = \dfrac{x - \mu}{\sigma}$

With SPSS, we can find these probabilities easily using the cumulative values you've calculated. First, let's take a look at the graph of the standard cumulative normal distribution.

🖰 **Graphs ➤ Line...** This time, choose a Simple line graph, still representing values of individual cases. The line should represent the cumulative probabilities for the variable with mean of 0 and standard deviation of 1. Choose Cum. Normal (0,1) [cn01]. Under Category labels, check Variable and select the variable x.

What is notable about the shape of CUMN01? Look at the point (0, 0.5) on this curve: What does it represent about the standard normal variable?

Now suppose we want to find $p(-2.5 < z < 1)$. We could scroll through cn01 to locate the probabilities, or we could request them directly, as follows.

🖰 In the empty variable column called Value, type in just the two numbers −2.5 and 1.

🖰 **Transform ➤ Compute...** For your Target Variable, type in cumprob. The Numeric Expression is CDF.NORMAL(value, 0,1).

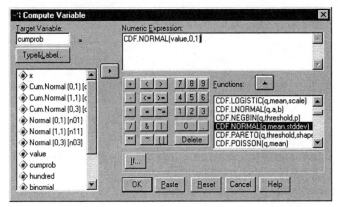

🖰 After clicking **OK**, you'll see this in the Data Editor:

n03	value	cumprob	hundred
.	-2.50	.0062	1
.0008	1.00	.8413	2

To find $p(-2.5 < z < 1)$, we subtract $p(-2.5 < z)$ from $p(z < 1)$. In other words, we compute $.8413 - .0062$, and get $.8351$. This approach works for any normally distributed random variable. Suppose that x is normal with a mean of 500 and a standard deviation of 100. Let's find $p(500 < x < 600)$.

🖰 Type 500 and 600 into the top two cells of value.

🖰 Edit the Compute dialog box again. In the Numeric Expression, change the mean parameter to 500, the standard deviation to 100. Once again, subtract the two cumulative probability values.

What is $p(500 < x < 600)$? $p(x > 600)$? $p(x < 300)$?

Normal Curves as Models

One reason the normal distribution is so important is that it can serve as a close approximation to a variety of other distributions. For example, binomial experiments with many trials are approximately normal. Let's try an example of a binomial variable with 100 trials, and $p(\text{success}) = .20$.

🖰 **Transform ➤ Compute...** The target variable is binomial, and the expression is CDF.BINOM (hundred, 100, .2). As in the prior examples, this generates the cumulative binomial distribution.

🖰 **Graphs ➤ Line...** Make a simple line graph representing the variable called binomial, with the values from hundred serving as labels for the category axis.

Do you see that this distribution could be approximated by a normal distribution? The question is, *which* normal distribution in particular? Since $n = 100$ and $p = .20$, the mean and standard deviation of the binomial variable are 20 and 4.[5] Let's generate a normal curve with those parameters.

🖰 **Transform ➤ Compute...** The target variable here is cn204, and the expression is CDF.NORMAL(hundred,20,4).

---

[5] For a binomial x, $E(x) = \mu = np$. Here, that's $(100)(.20) = 20$. The standard deviation is $\sigma = \sqrt{np(1-p)} = \sqrt{(100)(.20)(.80)} = \sqrt{16} = 4$.

🖱 **Graphs ➤ Line...** Create a multiple line plot representing individual cases. The two variables now are binomial and cn204, and the category axis is once again hundred. ***Would you say the two curves are approximately the same?***

The normal curve is often a good approximation of real-world observed data. Let's consider two examples.

🖱 Open the file **PAWorld**, which contains annual economic and demographic data from 42 countries.

🖱 **Graphs ➤ Histogram...** Select the variable Real consumption % of GDP [c]. Check the box labeled Display normal curve.

🖱 Do the same for the variable Per capita GDP relative to USA [y].

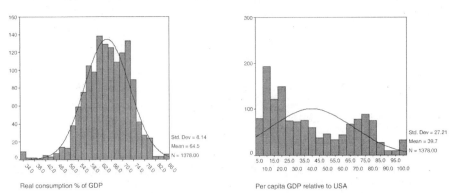

Real consumption % of GDP   Per capita GDP relative to USA

Does a normal distribution approximate either of these histograms? In your judgment, how closely does the normal curve approximate each histogram?

Moving On...

Normal (containing the simulated data from the first part of the session)

1. Use what you have learned to compute the following probabilities for a normal random variable with a mean of 8 and a standard deviation of 2.5:
   - $p(7 < x < 8.5)$
   - $p(9 < x < 10)$
   - $p(x > 4)$
   - $p(x < 4)$
   - $p(x > 10)$

2. Use the variables hundred and binomial to do the following: Generate cumulative probabilities for a binomial distribution with parameters $n = 100$ and $p = 0.4$. As illustrated in the session, also compute the appropriate cumulative normal probabilities (you must determine the proper μ and σ). Construct a graph to compare the binomial and normal probabilities; comment on the comparison.

Output

This file contains monthly data about the industrial output of the United States for many years. The first column contains the date, and the next six contain specific variables described in Appendix A. Generate six histograms with normal curves superimposed for *all six variables*.

3. Based on their histograms, which of the six variables looks most nearly normally distributed to you? Least nearly normal?

4. Suggest some real-world reasons that the variable you selected as most nearly normal would follow a normal distribution. That is, what characteristics of the particular variable could explain *why* it follows a normal curve?

BP

This file contains blood pressure readings and other measurements of a sample of individuals, under different physical and psychological stresses.

5. The variable dbprest refers to the resting diastolic blood pressure of the individuals. Generate a histogram of this variable, and comment on the extent to which it appears to be normally distributed.

6. Find the sample mean and standard deviation of dbprest. Use the CDF.NORMAL function and the sample mean and standard deviation to compute the probability that a randomly chosen person has a diastolic blood pressure in excess of 76.6. In other words, find $p(x > 76.6)$.

    In the sample, about 10% of the people had diastolic readings above 76.6. How does this compare to the normal probability you just found?

Bodyfat

This file contains body measurements of 252 men. Using the same technique described for the **Output** dataset, investigate these variables:

- FatPerc
- Age
- Weight
- Neck
- Biceps

7. Based on their histograms, which variable looks most nearly normally distributed to you? Least nearly normal?

8. Suggest some real-world reasons that the variable you selected as most nearly normal would follow a normal distribution.

9. For the neck measurement variable, find the sample mean and standard deviation. Use these values as the parameters of a normal curve, and generate the theoretical cumulative probabilities. Using these probabilities, *estimate* the percentage of men with neck measurements between 29 and 35 cm. In fact, 23 of the men in the sample (9.1%) did fall in

that range; how does this result compare to your estimate? Comment on the comparison.

Water

These data concern water usage in 221 regional water districts in the United States for 1985 and 1990. Compare the normal distribution as a model for Total freshwater consumptive use 1985 [tocufr85] and Consumptive use % of total use [pctcu85]. (You investigated these variables earlier in Session 4.)

10. Which one is more closely modeled as a normal variable?

11. What are the parameters of the normal distribution which closely fits the variable Consumptive use % of total use [pctcu85]?

12. What concerns, if any, might you have in modeling Consumptive use % with a normal curve? (*Hint*: Think about the range of possible values for a normal curve.)

MFT

This worksheet holds scores of 137 students on a Major Field Test (MFT), as well as their GPAs and SAT verbal and math scores.

13. Identify the parameters of a normal distribution which closely approximates the math scores of these students.

14. Use the mean and standard deviation of the distribution you have identified to *estimate* the proportion of students scoring above 59 on the math SAT.

15. In this sample, the third quartile (75th percentile) for math was 59. How can we reconcile your previous answer and this information?

Milgram

This dataset contains results of Milgram's famous experiments on obedience to authority. Under a variety of experimental conditions, subjects were instructed to administer electrical shocks to another person; in reality, there were no electrical shocks, but subjects believed that there were.

16. Create a histogram of the variable Volts. Discuss the extent to which this variable appears to be normally distributed. Comment on noteworthy features of this graph.

Sampling Distributions

Objectives

In this session, you will learn to do the following:
- Simulate random sampling from a known population
- Transfer output from the Viewer to the Data Editor
- Use simulation to illustrate the Central Limit Theorem

What Is a Sampling Distribution?

Every random variable has a probability distribution or a probability density function. One special class of random variables is *statistics computed from random samples.*

How can a statistic be a random variable? Consider a statistic such as the sample mean, $\bar{x}$. In a particular sample, $\bar{x}$ depends on the n values in the sample; a different sample would potentially have different values, probably resulting in a different mean. Thus, $\bar{x}$ is a quantity that varies from sample to sample, due to the chance process of random sampling. In other words, it's a quantitative random variable.

Every random variable has a distribution with shape, center, and spread. The term *sampling distribution* refers to the distribution of a sample statistic. In this lab session, we'll simulate drawing many random samples from populations whose distributions are known, and see how the sample statistics vary from sample to sample.

Sampling from a Normal Population

We start by simulating a large sample from a population known to be normally distributed, with $\mu = 500$ and $\sigma = 100$. We could use the menu commands repeatedly to compute random data. In this instance, it is more convenient to run a small program than do the repetitive work ourselves. In SPSS, we can use programs that are stored in *syntax files*.

This particular syntax file simulates drawing 100 random samples from this known population. When we do simulations with SPSS, we will want to start the process by *seeding* the internal random number generator. The *seed* value provides a starting point for the generator.

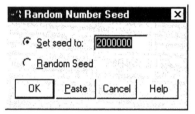

🖱 **Transform ➤ Random Number Seed...** Enter any number between 1 and 2,000,000,000 in this dialog. By default, SPSS uses 2,000,000. You should enter a different number.

🖱 **File ➤ Open ➤ Syntax...** In the Look in box, choose the directory you always select. Notice that the Files of type: box now says Syntax (*.sps). You should see three file names listed. Then select and open the syntax file called **Normgen**.

After opening the syntax file, you will see the Syntax Editor, which displays the program statements. Within the Syntax Editor window, do the following:

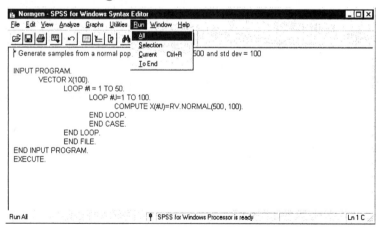

🖱 **Run ➤ All** This will execute the program, generating 100 columns of 50 observations each. In other words, we are simulating 100 different random samples of size $n = 50$, drawn from a normally distributed population whose mean is 500 and standard deviation is 100.

Close the Syntax Editor window. In the Data Editor, look at x1, x2, and x3. Remember that these are simulated random samples, different from one another and from your neighbors' and different still from the other samples you have generated. The question is, how *much* different?

Since the mean of the population is 500, it is reasonable to expect the mean of the first column to be near 500. It may or may not be "very" close, but the result of one simulation doesn't tell us much. To get a feel for the randomness of $\bar{x}$, we need to consider many samples. That's why this program generates 100 samples.

🖱 **Analyze ➤ Descriptive Statistics ➤ Descriptives...** Select all of the *x* variables as the variables to analyze.

After issuing this command, you'll see the results in the Output Viewer. Since each set of simulations is unique, your results will differ from those shown below. In the output, the column labeled Mean contains the sample means of all 100 samples. We could consider this list of means itself a random variable, since each sample mean is different due to the chance involved in sampling. ***What should the mean of all of these sample means be? Explain your rationale.***

🖱 In the Viewer window, double-click on the area titled Descriptive Statistics. Then, as you would in a word-processing document, click on the first value in the Mean column to select it.

🖱 Use the scroll bars to scroll down until you see the mean of X100; hold the Shift key on the keyboard, and click the left mouse button again. This should highlight the entire column of numbers, as shown on the next page.

Descriptive Statistics

	N	Minimum	Maximum	Mean	Std. Deviation
X1	50	303.99	731.64	480.0696	96.1023
X2	50	278.70	768.18	510.0833	117.2996
X3	50	300.99	709.25	477.9638	89.4562
X4	50	226.20	678.23	494.1271	86.8440
X5	50	239.23	691.56	487.9519	93.9286
X6	50	246.63	659.54	517.4884	99.0609
X7	50	295.61	738.56	506.7308	103.8892
X8	50	209.23	678.78	512.0811	90.5681
X9	50	287.51	688.58	483.4456	91.6514
X10	50	313.29	700.73	493.8483	83.6457
X11	50	295.48	869.60	508.4727	106.2462
X12	50	300.30	674.67	507.0333	97.0330

🖱 **Edit ➤ Copy** This will copy the list of sample means.

🖱 Switch to the Data Editor, and click on the Variable View tab. Scroll down to row 100, and name a new variable Means. You may keep all of the default settings for the new variable.

🖱 Click on the Data View tab, and scroll to the right to the first empty column, adjacent to x100. Then move the cursor into the first cell of the column and click once.

🖱 **Edit ➤ Paste** This will paste all of the 100 sample means into the column. Now you have a variable that represents the sample means of your 100 random samples.

Despite the fact that your random samples are unique and individually unpredictable, we can predict that the mean of Means will be very nearly 500. This is a key reason that we study sampling distributions. We can make very specific predictions about the sample mean in repeated sampling, even though we cannot do so for one sample.

How much do the sample means vary around 500? Recall that in a random sample from an infinite population, *the standard error of the mean* is given by this formula:

$$\sigma_{\bar{x}} = \frac{\sigma}{\sqrt{n}}$$

In this case, $\sigma = 100$ and $n = 50$. So here,

$$\sigma_{\bar{x}} = \frac{100}{\sqrt{50}} = \frac{100}{7.071} = 14.14$$

Let's evaluate the center, shape, and spread of the variable called Means. If the formula above is true, we should find that the standard deviation of Means is approximately 14.

🖱 **Graphs ➤ Histogram...** Select Means as the variable, and click on Display normal curve in the dialog box. Below is a histogram from one simulation (yours will look somewhat different).

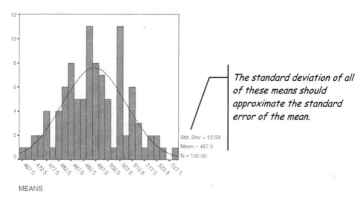

The standard deviation of all of these means should approximate the standard error of the mean.

MEANS

Notice the overall (but imperfect) bell shape of the histogram; the mean is so close to 500, and the standard deviation is approximately 14. Remember that the standard error is the theoretical standard deviation of all possible values of $\bar{x}$ and the standard deviation of Means represents only 100 of those samples.

How does your histogram compare to this one? What do you notice about their respective centers and spread?

Construct a histogram for any one of the x variables (x1 to x100). *Comment on the center, shape, and spread of this distribution, in comparison to the ones just discussed.*

Central Limit Theorem

The histogram of Means was roughly normal, describing the means of many samples from a normal population. That may seem reasonable—the means of samples from a normal population are themselves normal. But what about samples from non-normal populations? According to the Central Limit Theorem, the distribution of sample means approaches a normal curve as n grows large, *regardless of the shape of the parent population*.

To illustrate, let's take 100 samples from a uniform population ranging from 0 to 100. In a uniform population with a minimum value of a and a maximum value of b, the mean is found by:

$$E(x) = \mu = \frac{(a+b)}{2}$$

In this population, that works out to a mean value of 50. Furthermore, the variance of a uniform population is:

$$Var(x) = \sigma^2 = \frac{(b-a)^2}{12}$$

In this population, the variance is 833.33, and therefore the standard deviation is $\sigma = 28.8675$. Our samples will again have $n = 50$; according to the Central Limit Theorem, the standard error of the mean in such samples will be $28.8675/\sqrt{50} = 4.08$. Thus, the Central Limit Theorem predicts that the means of all possible 50 observation samples from this population will follow a normal distribution whose mean is 50 and standard error is 4.08. Let's see how well the theorem predicts the results of this simulated experiment.

🖰 **File ➤ Open ➤ Syntax...** This time open the file called **Unigen**. This syntax file generates 100 random samples from a uniform population like the one just described.

🖰 **Run ➤ All** Switch to the Data Editor, and notice that it now displays new values, all between 0 and 100.

🖰 **Analyze ➤ Descriptive Statistics ➤ Descriptives...** Select all of the x variables, and click **OK**.

🖰 As you did earlier, in the Output Viewer, select and copy all values in the Mean column, and paste them into a new variable called Means.

Once more, create a histogram for any x variable, and another histogram for the Means variable. As before, the reported "Std. Dev." should approximate the theoretical standard error of the mean. The results of one simulation are shown here:

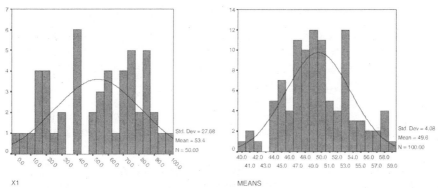

X1

MEANS

To what extent do the mean and standard error of Means approximate the theoretical values predicted by the Central Limit Theorem? Which graph of yours appears to be more closely normal?

Look closely at your two graphs (ours are shown above). ***What similarities do you see between your graphs and these? What differences? How do you explain the similarities and differences?***

Sampling Distribution of the Proportion

The examples thus far have simulated samples of a quantitative random variable. Not all variables are quantitative. The Central Limit Theorem and the concept of a sampling distribution also apply to qualitative random variables, with three differences. First, we are not concerned with the mean of the random variable, but with the *proportion (p)* of times that a particular outcome is observed. Second, we need to change our working definition of a "large sample." The standard guideline is that n is considered large if both $n \cdot p > 5$ and $n(1 - p) > 5$. Third, the formula for the standard error becomes:

$$\sigma_{\bar{p}} = \sqrt{\frac{p(1-p)}{n}}$$

To illustrate, we'll generate more random data. Recall what you learned about binomial experiments as a series of n independent trials of a process generating success or failure with constant probability, p, of success. Such a process is known as a *Bernoulli trial*. We'll construct 100 more samples, each consisting of 50 Bernoulli trials:

🖱 As in the prior two simulations, we'll run a syntax file. This time, the file is called **Berngen**. Open the file and run it.

This creates 100 columns of 0s and 1s, where 1 represents a success. By finding the mean of each column, we'll be calculating the relative frequency of successes in each of our simulated samples, also known as the *sample proportion, $\bar{p}$* .

🖱 Also as before, compute the descriptive statistics on the 100 samples, and then copy and paste the variable means into a newly created variable called Means.

Now Means contains 100 sample proportions. According to the Central Limit Theorem, they should follow an approximate normal distribution with a mean of 0.3, and a standard error of

$$\sigma_{\bar{p}} = \sqrt{\frac{p(1-p)}{n}} = \sqrt{\frac{(.3)(.7)}{50}} = .0648$$

As we have in each of the simulations, graph the descriptive statistics for one of the *x* variables and for Means. **Comment on the graphs you see on your screen** (ours are shown here).

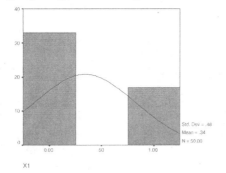

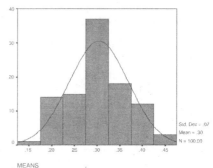

Moving On...

1. What happens when *n* is under 30? Does the Central Limit Theorem work for small samples too? Open **Unigen** again. In the Syntax Editor, find the command line that says

   LOOP #I = 1 TO 50.

Change the 50 to 20 (to create samples of $n = 20$), and run the program again. Compute the sample means, and create a histogram of the sample means. Close, but do not save, **Unigen**. Comment on what you see.

2. Open **Unigen**. Make the following changes to simulate samples from a uniform distribution ranging from –10 to 10.

   Edit the Compute line of the file to read as follows:

   COMPUTE X(#j)=RV.UNIFORM(-10,10)

   Report on the distribution of sample means from 100 samples of $n = 50$.

Pennies

This file contains the results of 1,685 repeated binomial experiments, and each consisted of flipping a penny 10 times. We can think of each 10-flip repetition as a sample of $n = 10$ flips; this file summarizes 1,685 different samples.

In all, nearly 17,000 individual coin flips are represented in the file. Each column represents a different possible number of heads in the 10-flip experiment, and each row contains the results of one student's repetitions of the 10-flip experiment. Obviously, the average number of heads should be 5, since the theoretical proportion is $p = 0.5$.

3. According to the formula for the standard error of the sample proportion, what should the standard error be in this case (use $n = 10$, $p = .5$)?

4. (Hint: For help with this question, refer to Session 8 for instructions on computing normal probabilities, or consult a normal probability table in your textbook.) Assuming a normal distribution, with a mean = 0.5 and a standard error equal to your answer to #3, what is the probability that a random sample of $n = 10$ flips will have a sample proportion of *0.25 or less*? (i.e., 2 or fewer heads)

5. Use the Frequencies statistics commands (see the Analyze menu) to determine whether these real-world penny data refute or support the predictions you made in your previous answer. What proportion of the samples contained 0, 1, or 2

heads respectively? Think very carefully as you analyze your SPSS output.

6. Comment on how well the Central Limit Theorem predicts the real-world results reported in your previous answer.

Colleges

Each of the colleges and universities in this *U.S. News and World Report* survey have been asked to submit the mean SAT scores of their freshman classes. Although many schools refused to provide this information, many schools did so. Thus, this is a sample of samples. It is generally assumed that SAT scores are normally distributed with mean 500 and standard deviation 100. For each of the following, comment about differences you notice and reasons they may occur.

7. Report on the distribution (center, shape, and spread) of the means for verbal SAT scores. Comment on the distribution.

8. Do the same for math scores. Comment on the distribution.

9. Repeat for combined SAT scores. Is there anything different about this distribution? Discuss.

Confidence Intervals

Objectives

In this session, you will learn to do the following:

- Construct large- and small-sample confidence intervals for a population mean
- Transpose columns and rows in SPSS output using Pivot Tables
- Construct a large-sample confidence interval for a population proportion

The Concept of a Confidence Interval

   A confidence interval is an estimate that reflects the uncertainty inherent in random sampling. To see what this means, we'll start by simulating random sampling from a hypothetical normal population, with $\mu = 500$ and $\sigma = 100$. Just as we did in the prior session, we'll create 100 simulated samples. Our goal is to learn something about the extent to which samples vary from one another.

   Transform ➤ Random Number Seed... Select any seed value between 1 and 2,000,000,000 as explained in Session 9.

   File ➤ Open ➤ Syntax... As you did in Session 9, find the syntax file called **Normgen**, and open it.

🖱 **Run ➤ All** This will simulate the process of selecting 100 random samples of size $n = 50$ observations, all drawn from a normally distributed population with $\mu = 500$ and $\sigma = 100$.

🖱 **Analyze ➤ Descriptive Statistics ➤ Explore...** This command will generate the confidence intervals. In the dialog, select all 100 of the *x* variables as the Dependent List, and click on the Display statistics radio button in the lower left.

In the Viewer window, you will see a Case Processing Summary, followed by a long Descriptives section. The layout of the Descriptives table makes it difficult to compare the confidence interval bounds for our samples. Fortunately, we can easily fix that by *pivoting* the table.

🖱 Double-click anywhere on the Descriptives section. This will open a Pivot Table window and a Pivoting Trays window. If you don't see the Pivoting Trays, select **Pivot ➤ Pivoting Trays**.

🖱 Move your cursor into the Pivoting Tray. Click and drag the Statistics pivot icon from the Row tray to the Column tray. This swaps the columns and rows in the table.

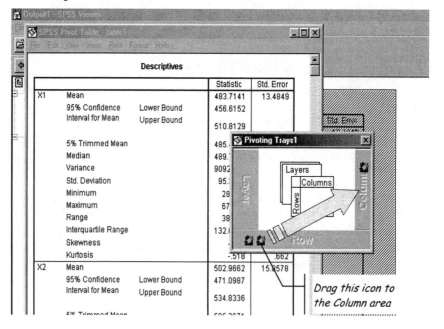

Now your Descriptives section should look more like the example shown below. Since these are simulated data, the specific values on your screen will differ from those shown.

| | | 95% Confidence Interval for Mean | | | | | | | |
	Mean	Lower Bound	Upper Bound	5% Trimmed Mean	Median	Variance	Std. Deviation	Minimum	Maximum
X1	480.0696	452.7576	507.3816	476.6217	473.8685	9235.657	96.1023	303.99	731.64
X2	510.0833	476.7471	543.4195	510.0358	491.9254	13759.207	117.2996	278.70	768.18
X3	477.9638	452.5406	503.3869	475.6140				99	709.25
X4	494.1271	469.4463	518.8079	495.5862				20	678.23
X5	487.9519	461.2577	514.6461	491.2456				23	691.56
X6	517.4884	489.3356	545.6412	522.6785				63	659.54
X7	506.7308	477.2058	536.2558	506.2734	492.0349	10792.905	103.8892	290.61	738.56
X8	512.0811	486.3420	537.8203	515.0435	527.0826	8202.589	90.5681	209.23	678.78
X9	483.4456	457.3986	509.4927	483.5371	489.1678	8399.982	91.6514	287.51	688.58
X10	493.8483	470.0765	517.6202	494.3301	501.1073	6996.601	83.6457	313.29	700.73
X11	508.4727	478.2778	538.6675	505.8184	515.9832	11288.262	106.2462	295.48	869.60
X12	507.0333	479.4597	534.6069	510.1789	519.8707	9413.421	97.0228	290.39	671.67
X13	497.6965	467.6612	527.7318	501.6050	503.2385	11169.317	105.6850	209.26	663.71
X14	506.9456	479.6488	534.2425	504.9093	517.0486	9225.436	96.0491	335.25	740.69
X15	512.0732	481.9984	542.1479	511.8740	508.8012	11198.649	105.8237	320.57	724.45
X16	481.4628	453.8645	509.0611	481.9665	497.6981	9430.310	97.1098	292.55	656.44
X17	473.3491	446.5008	500.1973	471.0597	468.5434	8924.700	94.4706	283.71	685.55
X18	499.7157	469.7182	529.7131	496.2994	491.3643	11141.126	105.5515	308.32	800.10
X19	502.3801	472.1858	532.5743	497.5571	487.0231	11287.831	106.2442	337.73	768.34
X20	485.0701	458.4475	511.6927	485.8877	495.7731	8775.319	93.6767	254.38	695.01
X21	469.7760	441.2377	498.3144	466.8899	471.8642	10083.667	100.4175	302.70	777.87
X22	512.3724	484.7198	540.0249	513.5178	510.1625	9467.404	97.3006	285.17	717.26

Descriptives / Statistic

These are the confidence intervals

For each of the 100 variables, there is one line of output, containing the variable name, sample mean, 95% confidence interval, and a number of other descriptive statistics.

In the sample output, one row is highlighted. In our simulation, the interval in this row lies entirely to the left of 500. Since this is a simulation, we know the true population mean ($\mu = 500$). Therefore, the confidence intervals ought to be in the neighborhood of 500. ***Do all of the intervals on your screen include 500? If some do not, how many don't?***

> Remember that in a simulation, each of us will generate 100 different samples, and have 100 different confidence intervals. In 95% interval estimation, about 5% (1 in 20) of all possible intervals don't include μ. Therefore, you should have approximately 95 "good" intervals.

Recall what you know about confidence intervals. When we refer to a 95% confidence interval we are saying that 95% of all possible random samples from a population would lead to an interval containing μ. Here you have generated merely 100 samples of the infinite number possible, but the pattern should become clear.

Effect of Confidence Coefficient

An important element of a confidence interval is the *confidence coefficient,* reflecting our degree of certainty about the estimate. By default, SPSS sets the confidence interval level at 95%, but we can change that value. Generally, these coefficients are conventionally set at levels of 90%, 95%, 98%, or 99%. Let's focus on the impact of the confidence coefficient by reconstructing a series of intervals for the first simulated sample.

🖱 Look at the Descriptives output on your screen, and write down the 95% confidence interval limits corresponding to sample x1.

🖱 **Analyze ➤ Descriptive Statistics ➤ Explore...** In the Dependent List, deselect all of the variables except x1. Click the button marked **Statistics....** In the dialog box (see below), change the 95% to 90, and click **Continue...** in the dialog box, and then **OK** in the Explore dialog box. *How do the 90% intervals compare to the 95% intervals?*

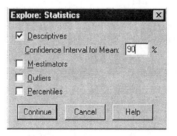

🖱 Do the same twice more, with confidence levels of 98% and 99%.

How do the intervals compare to one another? What is the difference from one interval to the next?

Large Samples from a Non-normal (Known) Population

Recall Session 9. We generated some large samples from a uniformly distributed population with a minimum value of 0 and a maximum of 100. In that session (see page 96), we computed that such a population has a mean of 50 and a standard deviation of 28.8675.

According to the Central Limit Theorem, the means of samples drawn from such a population will approach a normal distribution with a mean of 50 and a standard error of $28.8675/\sqrt{n}$ as n grows large. For most practical purposes, when n exceeds 30 the distribution is approximately normal; with a sample size of 50 we should be comfortably in the "large" range. As we did in the previous session, we will simulate 100 random samples of 50 cases each.

🖰 **File ➤ Open ➤ Syntax...** Open and run the file called **Unigen**.

🖰 **Analyze ➤ Descriptive Statistics ➤ Explore...** Select all 100 columns, set the confidence interval level to 95% once again (by clicking on **Statistics**), and create the confidence intervals.

🖰 Pivot the Descriptives table as we did earlier.

Again, review the output looking for any intervals that exclude 50. ***Do we still have about 95% success? How many of your intervals exclude the true mean value of 50?***

Dealing with Real Data

Perhaps you now have a clearer understanding of a confidence interval. It is time to leave simulations behind us, and enter the realm of real data where we don't know μ or σ. For large samples (usually meaning $n > 30$), the traditional "by-hand" approach is to invoke the Central Limit Theorem, to estimate σ using the sample standard deviation (s), and to construct an interval using the normal distribution. With software like SPSS, the default presumption is that we don't know σ, and so the **Explore** command automatically uses the sample standard deviation and builds an interval using the values of the *t distribution*[1] rather than the normal.

Even with large samples, we should use the normal curve only when σ is known—which very rarely occurs with real data. Otherwise, the *t* distribution is appropriate. In practice, the values of the normal and *t* distributions become very close when n exceeds 30. With small samples, though, we face different challenges.

---

[1] The *t* distribution is a family of bell-shaped distributions. Each *t* distribution has one parameter, known as *degrees of freedom (df)*. In the case of a single random variable, $df = n-1$. See your primary text for further information.

Small Samples from a Normal Population

If a population cannot be assumed normal, we must use large samples or nonparametric techniques such as those presented in Session 21. However, if we can assume that the parent population is normal, then small samples can be handled using the *t* distribution. Let's take a small sample from a population which happens to be normal: SAT scores of incoming college freshmen.

🖰 **File ➤ Open ➤ Data...** Select **Colleges**.

🖰 **Analyze ➤ Descriptive Statistics ➤ Explore...** Select the variable Avg Combined SAT [combsat] as the only variable in the Dependent List.

🖰 Before clicking **OK,** under Display, be sure that Both is selected. Then click on the **Plots...** button to open the dialog box shown below. Complete it as shown, and then click **Continue** and **OK.**

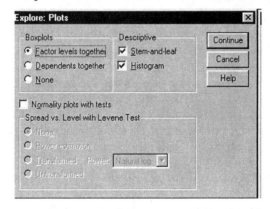

In the Viewer window, we first note a substantial number of missing observations; in this dataset, many schools did not report mean SAT scores. Before looking at the interval estimates, first scroll down and look at the histogram for the variable. It strongly suggests that the underlying variable is normally distributed. From this output, we can also find the mean and standard deviation. Let's treat this dataset as a population of U.S. colleges and universities, and use it to illustrate a small-sample procedure. In this population, we know that $\mu = 967.98$ and $\sigma = 123.58$, and the population is at least roughly normal.

To illustrate how we would analyze a small sample, let's select a small random sample from it. We'll use the sample mean to construct a confidence interval for μ. Switch to the Data Editor.

🖱 **Data ➤ Select Cases...** Select Random sample of cases, and click the button marked **Sample....** Specify that we want exactly 30 cases from the first 1,302 cases, as shown below. Since roughly 60% of the schools reported mean SAT scores, this should give us about 18 cases to work with in our sample.

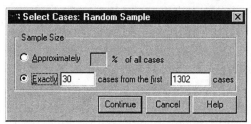

🖱 **Analyze ➤ Descriptive Statistics ➤ Explore...** Look at the resulting interval in your Viewer window (our output appears here). ***Does it contain the actual value of μ? Will everyone in the class see an interval containing μ? Explain.***

Case Processing Summary

	Cases					
	Valid		Missing		Total	
	N	Percent	N	Percent	N	Percent
Avg Combined SAT	19	63.3%	11	36.7%	30	100.0%

Descriptives

			Statistic	Std. Error
Avg Combined SAT	Mean		945.05	33.73
	95% Confidence Interval for Mean	Lower Bound	874.18	
		Upper Bound	1015.92	
	5% Trimmed Mean		940.23	
	Median		926.00	
	Variance		21621.164	
	Std. Deviation		147.04	
	Minimum		703	
	Maximum		1274	
	Range		571	
	Interquartile Range		204.00	
	Skewness		.610	.524
	Kurtosis		.009	1.014

Moving On...

Colleges

1. With the full dataset, construct a 95% confidence interval estimate for the mean room and board charges at U.S. colleges.

2. Does this interval indicate that 95% of all students in America have room and board costs in this interval? Explain.

F500

3. Construct a 95% confidence interval for mean Profit to Sales Ratio. What does this interval tell us?

4. Can we consider the 1996 Fortune 500 a random sample? What would the parent population be?

5. Does this variable appear to be drawn from a normal population? What evidence would you consider to determine this?

Swimmer

This file contains the times for a team of high school swimmers in various events. Each student recorded two "heats" or trials in at least one event.

6. Construct a 90% confidence interval for the mean of first times in the 100-meter freestyle. (Hint: Use eventrep as a factor; you'll need to read the output selectively to find the answer to this question).

7. Do the same for the second times in the 100-meter freestyle.

8. Comment on the comparison of the two intervals you've just constructed. Suggest real-world reasons which might underlie the comparisons.

Eximport

This file contains monthly data about the dollar value of U.S. exports and imports for the years 1948–1996. Consult Appendix A for variable identifications.

9. Estimate the mean value of exports to the United States, excluding military aid shipments. Use a confidence level of 95%.

10. Estimate the mean value of General Imports, also using a 95% confidence level.

11. On average, would you say that the United States tends to import more than it exports (excluding military aid shipments)? Explain, referring to your answers to #9 and #10.

12. Estimate the mean value of imported automobiles and parts for the period covered in this file, again using a 95% confidence level.

MFT

This data is collected from students taking a Major Field Test (MFT). Students' SAT scores are also included.

13. Construct 95% confidence intervals for both verbal and math SAT scores. Comment on what you find. Knowing that SAT scores nationally have a mean theoretical score of 500, are these intervals what you might expect to see? Based on these intervals, would you infer that this was a major field test in the natural sciences or in the humanities? Explain your reasoning.

14. Construct a 95% confidence interval for the mean score of total MFT. Comment on your findings.

Sleep

This data is a collection of sleep habits of mammals. It also includes life expectancies for each animal. Note that humans sleep an average of 8 hours a day and that maximum human life expectancy is 100 years.

15. Construct 95% confidence intervals for both total sleep and life span. Comment on anything interesting you notice. How might life span and total sleep relate to each other?

Session 11

One-Sample Hypothesis Tests

Objectives

In this session, you will learn to do the following:
- Perform hypothesis tests concerning a population mean
- Verify conditions for a small-sample test

The Logic of Hypothesis Testing

In the previous lab, the central questions involved estimating a population parameter—questions such as, "What is the value of μ?" In many instances, we are less concerned with estimating a parameter than we are with comparing it to a particular value—that is questions such as, "Is μ more than 7?" This session investigates this second kind of question. To underscore the distinction between these two kinds of questions, consider an analogy from the justice system. When a crime has been committed, the question police ask is, "Who did this?" Once a suspect has been arrested and brought to trial, the question for the jury is, "Did the defendant do this?" Although the questions are clearly related, they are different and the methods for interpreting the available evidence are also different.

Random samples provide evidence about a population. In a hypothesis test, we generally have an initial presumption about the population, much like having a defendant in court. The methods of hypothesis testing are designed to take a cautious approach to the weighing of such evidence. The tests are set up to give substantial advantage to the initial belief, and only if the sample data are very compelling do we abandon our initial position. In short, the methods of hypothesis testing provide a working definition of *compelling evidence*.

In any test, we start with a *null hypothesis,* which is a statement concerning the value of a population parameter. We could, for example, express a null hypothesis as follows: "At least 75% of news coverage is positive in tone," or "The mean weekly grocery bill for a family in our city is $150." In either case, the null hypothesis states a presumed value of the parameter. The purpose of the test is to decide whether data from a particular sample are so far at odds with that null hypothesis as to force us to reject it in favor of an *alternative* hypothesis.

An Artificial Example

We start with some tests concerning a population mean, and return to the first simulation we conducted in introducing confidence intervals. In that case, we simulated drawing random samples from a normal population with $\mu = 500$ and $\sigma = 100$. We'll do the same thing again, understanding that there are an infinite number of possible random samples, each with its own sample mean. Though it is likely our samples will each have a sample mean of about 500, it is *possible* we will obtain a sample with a mean so far from 500 that we might be convinced the population mean is not 500.

What's the point of the simulation? Remember that this is an *artificial* example. Usually, we do not know the truth, and are trying to infer it from a random sample. Ordinarily, we would not know the truth about μ; we would have one sample, and we would be asking if this sample is consistent with the hypothesis that $\mu = 500$. This simulation can give us a feel for the risk of an incorrect inference based on any single sample. As in prior sessions, we first seed the random number generator.

🖱 **Transform ➤ Random Number Seed...** Set the seed to a whole number of your choice.

🖱 **File ➤ Open ➤ Syntax...** Open and run **Normgen.sps**. This will generate 100 pseudorandom samples from a population with $\mu = 500$ and $\sigma = 100$, as in previous sessions.

🖱 **Analyze ➤ Compare Means ➤ One-Sample T Test...** Select all 100 variables, enter 500 into the box labeled Test Value, and click **OK**.

This will perform a one-sample test of the null hypothesis that μ = 500, using each of our 100 samples.

A *t test* compares sample results to a hypothesized value of μ, using a *t* distribution as the standard of comparison. A *t* distribution is bell-shaped and has one parameter, known as *degrees of freedom* (*df*). In the one-sample *t* test, $df = n-1$.

Now look at Viewer window (a portion of our results is shown here). Remember that your output will be different since this is a simulation. First you'll see a table of sample means, standard deviations, and standard errors for each sample. Then, you'll see the test results:

One-Sample Test

					95% Confidence Interval of the Difference	
					Test Value = 500	
	t	df	Sig. (2-tailed)	Mean Difference	Lower	Upper
X1	-.579	49	.565	-9.4498	-42.2212	23.3216
X2	1.033	49	.307	15.3439	-14.5142	45.2019
X3	-.698	49	.489	-10.2861	-39.9020	19.3299
X4	-3.024	49	.004	-37.4215	-62.2924	-12.5505
X5	-.541	49	.591	-8.4276	-39.7187	22.8635
X6	1.103	49	.275	14.2557	-11.7071	40.2184
X7	.858	49	.395	10.1067	-13.5612	33.7746
X8	-.842	49	.404	-12.0659	-40.8672	16.7354
X9	1.535	49	.131	20.4249	-6.3234	47.1731
X10	-.687	49	.495	-9.9759	-39.1399	19.1882
X11	.408	49	.685	5.6530	-22.2010	33.5070
X12	.017	49	.987	.2463	-29.2551	29.7476
X13	-.814	49	.419	-11.9809	-41.5480	17.5861
X14	2.343	49	.023	27.9152	3.9735	51.8569
X15	.301	49	.764	4.2368	-24.0039	32.4775
X16	-1.392	49	.170	-18.2947	-44.7041	8.1146
X17	.831	49	.410	11.1818	-15.8722	38.2359
X18	1.004	49	.320	13.2472	-13.2757	39.7700
X19	.198	49	.844	2.7669	-25.3161	30.8500

The output reports the null hypotheses, and summarizes the results of these random samples. In this example, the sample mean of X1 (not shown) was 490.5502. This is below 500, but is it so far below as to cast serious doubt on the hypothesis that $\mu = 500$? The *test statistic* gives us a relative measure of the sample mean, so that we can judge how consistent it is with the null hypothesis. In a large-sample test with a normal population, the test statistic[1] is computed as follows:

$$t = \frac{\bar{x} - \mu}{s/\sqrt{n}} = \frac{490.5502 - 500}{115.3122/\sqrt{50}} = -.579$$

This is the value reported in the t column of the output. In other words, 490.5502 is only 0.579 standard errors below the hypothesized value of μ. Given what we know about normal curves, that's not very far off at all, and the same is true for this t distribution. It is quite consistent with the kinds of random samples one would expect from a population with a mean value of 500.

In fact, we could determine the likelihood of observing a sample mean more than 0.579 standard errors away from 500 in either direction. That likelihood is called the *P-value* and it appears in the column marked Sig. (2-tailed). In this particular instance, $P \le .565$. "Sig." refers to the smallest significance level (α) that would lead to rejection of the null hypothesis, and "2-tailed" indicates that this P-value is computed for the two-sided alternative hypothesis ($\ne$). If our alternative hypothesis were one-sided (> or <), and the sample results consistent with the alternative hypothesis, we would divide the reported P-value in half. In a one-sided test, if the sample result is at odds with the alternative, we compute the P-value as 1 – (Sig./2).[2]

A *statistically significant* result is one that is unlikely to have occurred by the chance involved in random sampling. It may or may not be practically important to the researcher, but the researcher can be confident that the finding is not due to sampling error. When we reject the null, we say that we have a statistically significant finding.

One way of thinking about the P-value is that if you were to reject the null hypothesis on the basis of this test, there is a probability of at

---

[1] In the rare event when σ is known, the test statistic is called z, and uses σ instead of s. SPSS does not provide a command to compute a z-test statistic, and by default uses s.

[2] An illustrative example might clarify this point. If our alternative hypothesis had been $\mu < 500$, then our sample statistic of 490.55 would be consistent with that alternative. In that event, $P = .565/2 = .283$. On the other hand, if the alternative hypothesis were $\mu > 500$, the sample mean would have been inconsistent with the alternative. In that case, we'd compute $P = 1-(5.65/32) = .718$.

most 0.565 that you are making a Type I error.[3] Since that probability is so high, you would be well advised against rejecting the null hypothesis in this instance.

Look down the list of test statistics and *P*-values in the output shown on page 113. Note that two lines are highlighted. In random sample x4, the test statistic was –3.024, and the *P*-value was only 0.004. For this sample, at a significance level of α = .05, we would *reject* the null hypothesis, and erroneously conclude that the population mean is not equal to 500. We would similarly have been misled by sample x14, but in all of the other simulated samples, we would have come to the correct conclusion.

Since this is a simulation, we know that the true population mean *is* 500. Consequently, we know that the null hypothesis really is true, and that most samples would reflect that fact. We also know that random sampling involves uncertainty, and that the population does have variation within it. Therefore, some samples (typically about 5%) will have *P*-values sufficiently small that we would actually reject the null hypothesis.[4]

What happened in your simulation? Assuming a desired significance level of α ***= .05, would you reject the null hypothesis based on any of these samples? What kinds of results do you think other people in the class generated?***

A More Realistic Case: We Don't Know Mu or Sigma

Simulations are instructive, but are obviously artificial. This simulation is unrealistic in at least two respects—in real studies, we generally don't know μ or σ, and we have only one sample to work with. Let's see what happens in a more realistic case.

 File ➤ Open ➤ Data... Open the **Bodyfat** file.

Each of us carries around different amounts of body fat. There is considerable evidence that important health consequences relate to the percentage of fat in one's total body mass. According to one popular

---

[3] A Type I error is rejecting the null hypothesis when it is actually true. Consult your textbook for further information about Type I errors and about *P*-values.

[4] In principle, α can be any value; in practice, most researchers and decision makers tend to use α = .05 or α = .01.

health and diet author, fat constitutes 23% of total body mass (on average) of adult males in the United States.[5]

Our data file contains body fat percentages for a sample of 252 males. Is this sample consistent with the assertion that the mean body fat percentage of the adult male population is 23%?

Since we have no reason to suspect otherwise, we can assume that the sample does come from a population whose mean is 23%, and establish the following null and alternative hypotheses for our test:

$$H_o: \quad \mu = 23$$
$$H_A: \quad \mu \neq 23$$

The null hypothesis is that this sample comes from a population whose mean is 23% body fat. The two-sided alternative is that the sample was drawn from a population whose mean is other than 23.

The dataset represents a large sample ($n = 252$), so we can rely on the Central Limit Theorem to assert that the sampling distribution is approximately normal, assuming this is a random sample. However, since we don't know σ, we should use the t test.

🖰 **Analyze ➤ Compare Means ➤ One-Sample T-Test...**   Select % body fat [fatperc], and enter the hypothetical mean value of 23 as the Test Value. The results are shown here:

T-Test

One-Sample Statistics

	N	Mean	Std. Deviation	Std. Error Mean
% body fat	252	19.1508	8.3687	.5272

One-Sample Test

	Test Value = 23					
					95% Confidence Interval of the Difference	
	t	df	Sig. (2-tailed)	Mean Difference	Lower	Upper
% body fat	-7.301	251	.000	-3.8492	-4.8875	-2.8109

Here the sample mean is 19.15% body fat, and the value of the test statistic, t, is a whopping –7.3 standard errors. That is to say, the

---

[5]Barry Sears, *The Zone* (New York: HarperCollins, 1995)

sample mean of 19.15% is extremely far from the hypothesized value of 23%. The *P*-value of approximately 0.000 suggests that we should confidently *reject* the null hypothesis, and conclude that these men are selected from a population with a mean body fat percentage of something other than 23.

In this example, we had a large sample and we didn't know σ. The Central Limit Theorem allowed us to assume normality and to conduct a *t* test. What happens when the sample is small? If we can assume that we are sampling from a normal population, we can perform a reliable *t* test even when the sample is small (generally meaning $n < 30$).

A Small-Sample Example

At the North East Region of the American Red Cross Blood Services, quality control is an important consideration. The Red Cross takes great pains to ensure the purity and integrity of the blood supply that they oversee. One dimension of quality control is the regular testing of blood donations. When volunteers donate blood or blood components, the Red Cross uses electronic analyzers to measure various components (white cells, red cells, etc.) of each donation.

One blood collection process is known as platelet pheresis. Platelets are small cells in blood that help to form clots, permitting wounds to stop bleeding and begin healing. The pheresis process works as follows: a specialized machine draws blood from a donor, strips the platelets from the whole blood, and then returns the blood to the donor. After about 90 minutes, the donor has provided a full *unit* of platelets for a patient in need. For medical use, a unit of platelets should contain about 4.0×10^{11} platelets.

The Red Cross uses machines made by different manufacturers to perform this process. Each machine has an electronic monitor that estimates the number of platelets in a donation. Since all people vary in their platelet counts, and since the machines vary slightly as well, the Red Cross independently measures each donation, analyzing the number of platelets in the unit.

Open the file called **Pheresis**, containing 294 readings from all donations in a single month.

One new machine in use at the Red Cross Donation Center is made by Amicus; in the sample, only sixteen donations were collected using the Amicus machine. Suppose that the Red Cross wants to know if the new machine is more effective in collecting platelets than one of the older models, which averages approximately 3.9 x 10^{11} platelets per unit. Initially, they assume that the new machine is equivalent to the old, and ask if the evidence strongly indicates that the new machine is more effective. The hypotheses are as follows:

H_0: $\mu \leq 3.9$ [new machine no better than old]
H_A: $\mu > 3.9$ [new machine more effective]

Note that this is a *one-sided* alternative hypothesis. Only sample means above 3.9 could possibly persuade us to reject the null. Therefore, when we evaluate the test results, we'll need to bear this in mind.

To isolate the Amicus donations, we do this:

Data ► Select Cases... Complete the dialog boxes as shown here to select only those cases where machine = 'Amicus'.

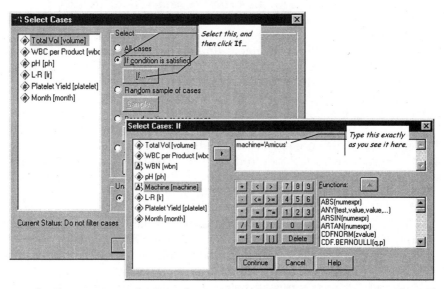

In this case, we have only sixteen observations, meaning that the Central Limit Theorem does not apply. With a small sample, we should only use the *t* test if we can reasonably assume that the parent population is normally distributed. Before proceeding to the test, therefore, we should look at the data to make a judgement about normality. The simplest way to do that is to make a histogram, and look for the characteristic bell shape.[6]

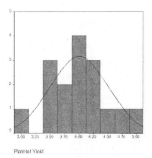

Platelet Yield

🖱 **Graphs ➤ Histogram...** Select the variable called Platelet Yield [platelet], and construct a histogram with a normal curve superimposed.

Look at the resulting histogram. It is generally mound-shaped, and therefore is acceptable for our purposes. Note that the histogram follows the normal curve relatively well.

That being the case, we can proceed with the t test. If the distribution were not bell-shaped, we would need to use a nonparametric technique, described in Session 21.

🖱 **Analyze ➤ Compare Means ➤ One-Sample T-Test...** In the dialog box, select the variable Platelets, specify a hypothesized value of 3.9, and click **OK**.

T-Test

One-Sample Statistics

	N	Mean	Std. Deviation	Std. Error Mean
Platelet Yield	16	4.0063	.5039	.1260

One-Sample Test

	Test Value = 3.9					
					95% Confidence Interval of the Difference	
	t	df	Sig. (2-tailed)	Mean Difference	Lower	Upper
Platelet Yield	.843	15	.412	.1062	-.1623	.3748

Remember that we must divide the reported significance level in half, since we want a 1-tailed test. ***On the basis of this test, what is your conclusion, assuming that α = .05? Is the mean platelet yield***

---

[6] Actually, there are statistical tests we can apply to decide whether or not a variable is normally distributed. Such tests will be introduced in Session 13.

greater than 3.9 x 10¹¹? Is the Amicus machine more efficient than the other machine?

Moving On...

Now apply what you have learned in this session. You can use one-sample *t* tests for each of these questions. Report on the relevant test results, explain what you conclude from each test, and why. Comment on the extent to which the normality assumption is satisfied for small samples.

GSS94

This is the subset of the 1994 General Social Survey.

1. Test the null hypothesis that adults in the United States watch an average of three hours of television daily.

PAWorld

This dataset contains multiple yearly observations for a large sample of countries around the globe.

2. One of the variables is called C, and represents the percentage of Gross Domestic Product (GDP) consumed within the country for the given year. The mean value of C in the sample is 64.525%. Was this average value *significantly* less than 65% of GDP? (In other words, would you reject a null hypothesis that $\mu \geq 65.0$ at a significance level of $\alpha = .05$?)

Bev

3. Test the null hypothesis that the mean current ratio for this entire sample of firms is equal to 3.0.

4. Using the **Select Cases** command, isolate the bottled and soft drink firms (SIC = 2086) and their current ratios. Use an appropriate test to see if the current ratio for these companies is *significantly different* from 3.0.

5. What about the malt beverage firms (SIC = 2082)? Is their ratio significantly different from 3.0?

BP

6. According to the World Health Organization, a normal, healthy adult should have a maximum systolic blood pressure of 140, and a maximum diastolic pressure of 90. Using the resting blood pressure readings from these subjects, test the hypothesis that their blood pressure readings are within the healthy range.

7. Many adults believe that a normal resting heart rate is 72 beats per minute. Did these subjects have a mean heart rate significantly different from 72 while performing a mental arithmetic task? Comment on what you find.

London2

These data were collected in West London and represent the hourly carbon monoxide (CO) concentration in the air (parts per million, or ppm) for the year 1996. For these questions, use the daily readings for the hour ending at 12 noon. You will perform one-sample t tests with this column of data.

8. In 1990, the first year of observations, West London had a mean carbon monoxide concentration of 1.5 ppm. One reason for the routine monitoring was the government's desire to reduce CO levels in the air. Is there a significant change in carbon monoxide concentration between 1990 and 1996? What does your answer tell you?

9. Across town at London Bexley, the 1996 mean carbon monoxide observation was .4 ppm. Is there a significant difference between London Bexley and West London?

10. What about London Bridge Place, which observed .8 ppm?

11. London Cromwell Road reported 1.4 ppm. Is their CO concentration significantly different than the concentration of CO in West London?

12. What do you think caused the differences (or lack thereof)?

Two-Sample Hypothesis Tests

Objectives

In this session, you will learn to do the following:

- Perform hypothesis tests concerning the difference in means of two populations
- Investigate assumptions required for small-sample tests
- Perform hypothesis tests concerning the difference in means of two "paired" samples drawn from a population

Working with Two Samples

In the prior lab session, we learned to make inferences about the mean of a population. Often our interest is in *comparing the means of two distinct populations.* To make such comparisons, we must select two independent samples, one from each population.

For samples to be considered independent, we must have no reason to believe that the observed values in one sample could affect or be affected by the observations in the other, or that the two sets of observations arise from some shared factor or influence.

We know enough about random sampling to predict that any two samples will likely have different sample means *even if they were drawn from the same population.* We anticipate some variation between any two sample means. Therefore, the key question in comparisons of samples from two populations is this: Is the observed difference between two sample means large enough to convince us that the populations have different means?

Sometimes, our analysis focuses on two distinct groups within a single population, such as female and male students. Our first example does just that. For starters, let's test the radical theory that male college students are taller than female college students. Open the **Student** file.

We can restate our theory in formal terms as follows:

$$H_o: \mu_f - \mu_m \geq 0$$
$$H_A: \mu_f - \mu_m < 0$$

Note that the hypothesis is expressed in terms of the *difference* between the means of the two groups.[1] The null says that men are no taller than women, and the alternative is that men are taller on average.

In an earlier session, we created histograms to compare the distribution of heights for these male and female students. At that time, we visually interpreted the graphs. We'll look at the histograms again, examine side-by-side boxplots of height, and then do the formal hypothesis test.

You may have learned that a two-sample t test requires three conditions:

- Independent samples
- Normal populations
- Equal population variances (for small samples)

The last item is not actually required to perform a t test. The computation of a test statistic is different when the variances are equal, and SPSS actually computes the test results for both equal and unequal variances, as we'll see. When variances are in fact unequal, treating them as equal may lead to a seriously flawed result.

As for the assumption of normality, the t test is reliable as long as the samples suggest symmetric, bell-shaped data without gross departures from a normal distribution. Since human height is generally normal for each sex, we should be safe here. Nevertheless, we are well-advised to examine our data for the bell shape.

> **Graphs ➤ Interactive ➤ Histogram...** By now, you should know your way around this dialog box. We want to place height on the horizontal axis and use gender as the *panel variable*. Superimpose a normal curve for these histograms by clicking

---

[1] When we establish hypotheses in a two-sample test, we arbitrarily choose the ordering of the two samples. Here, we express the difference as $\mu_f - \mu_m$ but we could also have chosen to write $\mu_m - \mu_f$.

on the Histogram tab and selecting Normal curve. Also, title your graph. When you have done so, you should see this:

Student Heights

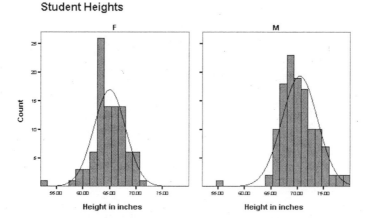

Prepared by R. Carver & J. Nash

Though not perfectly normal, these are reasonably symmetrical and bell-shaped, and suitable for performing the *t* test.[2] Before proceeding with the test to compare the mean height of the two groups, let's visualize these two samples using a boxplot:

🖰 **Graphs ➤ Interactive ➤ Boxplot...** Complete the dialog box as shown.

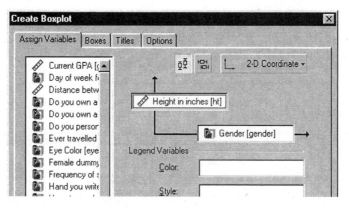

---

[2] Session 21, on nonparametric techniques, addresses the situation in which we have non-normal data.

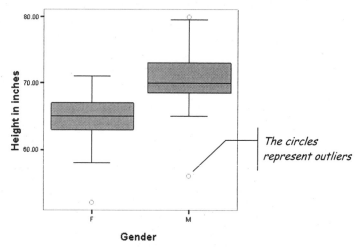

The circles represent outliers

As in the histogram, we see that both distributions are roughly symmetrical, with comparable variance or spread. The median lines in each box suggest that the averages are different; the test results will determine whether the extent of the difference is more than we would typically expect in sampling.

🖱 **Analyze ➤ Compare Means ➤ Independent-Samples T Test...**
Complete the dialog box as shown here, selecting Height in inches [ht] as the Test Variable, and Gender [gender] as the Grouping Variable.

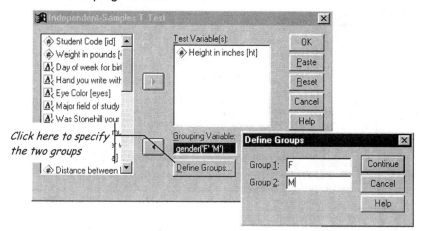

After clicking **OK**, you will see the test results in the Viewer window. As noted earlier, SPSS provides results both for the equal and unequal variance cases; it is up to us to select the appropriate result.

T-Test

Group Statistics

	Gender	N	Mean	Std. Deviation	Std. Error Mean
Height in inches	F	95	65.0316	2.9847	.3062
	M	123	70.4472	3.3753	.3043

Levene's test tells us whether to assume equal or unequal variances

Independent Samples Test

		Levene's Test for Equality of Variances		t-test for Equality of Means						95% Confidence Interval of the Difference	
		F	Sig.	t	df	Sig. (2-tailed)	Mean Difference	Std. Error Difference	Lower	Upper
Height in inches	Equal variances assumed	1.708	.193	-12.347	216	.000	-5.4156	.4386	-6.2801	-4.5511
	Equal variances not assumed			-12.544	212.023	.000	-5.4156	.4317	-6.2666	-4.5645

Recall that we have a one-sided less-than alternative hypothesis: The mean height for women is hypothesized to be less than the mean height for men. Therefore, evidence consistent with the alternative should lead to a negative difference. We notice here that, indeed, the women's sample mean is less than the men's. Also recall that the computations in the two-sample t test are performed differently for equal and unequal population variances.

Look at the output marked "Independent Samples Test." First look at the two blocks titled "Levene's Test for Equality of Variances." This is another of many statistical tests; the null hypothesis in Levene's test is that the variances of the two populations are equal. The test statistic, F, has a value of 1.708 and a P-value (significance) of .193.[3] In any statistical test, when P is less than our α, we reject the null. Here, with a large P-value, we do not reject the null, meaning that we can assume the variances to be equal.

To the right of the Levene test results, there are two rows of output for the variable, corresponding to equal and unequal variance conditions. Since we can assume equal variances for this test, we'll read only the top line. We interpret the output much in the same way as in the one-sample test. The test statistic t equals –12.347. The sign of the test statistic is what we expect if the alternative hypothesis is true. SPSS gives us a 2-tailed P-value, but this particular test is a 1-tailed test. Since the P-value here is approximately 0, the 1-tailed P-value is also

---

[3] Session 13 introduces other F-based tests.

about 0. We would reject the null hypothesis in favor of the alternative, and conclude the mean height for females is less than that for males.

Paired vs. Independent Samples

In the prior examples, we have focused on differences inferred from two independent samples. Sometimes, though, our concern is with the *change* in a single variable observed at two points in time. For example, to evaluate the effectiveness of a weight-loss clinic with fifty clients, we need to assess the change experienced by each individual, and not merely the collective gain or loss of the whole group.

We could regard such a situation as an instance involving two different samples. These are sometimes called *matched samples*, *repeated measures*, or *paired observations*.[4] Since the subjects in the samples are the same, we pair the observations for each subject in the sample, and focus on the difference between two successive observations or measurements. The only assumption required for this test is that the differences be normally distributed.

🖰 Open the file called **Swimmer2**.

This file contains swimming meet results for a high school swim team. Each swimmer competed in one or more different events, and for each event, the file contains the time for the swimmer's first and last "heat," or trial in that event. The coach wants to know if swimmers in the 100-meter freestyle event improve between the first and last time they compete. In other words, he wants to see if their second times are faster (lower) than their first times.

Specifically, the coach is interested in the *difference* between first and second times for each student. We are performing a test to see if there is evidence to suggest that times decreased; that is the suspicion that led us to perform the test. Therefore, in this test, our null hypothesis is to the contrary, which is that there was no change. Let μ_D equal the mean difference between the first and second times.

H_0: $\mu_D \leq 0$ [times did not diminish]
H_A: $\mu_D > 0$ [times did diminish]

---

[4] Paired samples are not restricted to "before-and-after" studies. Your text will provide other instances of their use. The goal here is merely to illustrate the technique in one common setting.

The fact that we repeatedly observed each swimmer means that the samples are not independent. Presumably, a student's second time is related to her first time. We treat these paired observations differently than we did with independent samples.

🖱 **Analyze ➤ Compare Means ➤ Paired-Sample T Test...** Choose the two variables related to the 100-meter freestyle, as shown here, and click **OK**.

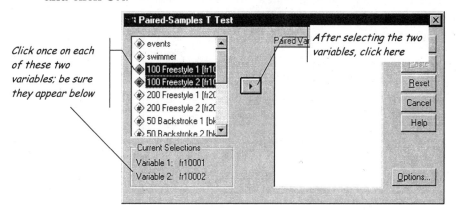

T-Test

Paired Samples Statistics

		Mean	N	Std. Deviation	Std. Error Mean
Pair 1	100 Freestyle 1	76.1072	32	10.1949	1.8022
	100 Freestyle 2	70.1141	32	9.0255	1.5955

Paired Samples Correlations

		N	Correlation	Sig.
Pair 1	100 Freestyle 1 & 100 Freestyle 2	32	.735	.000

Paired Samples Test

		Paired Differences							
					95% Confidence Interval of the Difference				
		Mean	Std. Deviation	Std. Error Mean	Lower	Upper	t	df	Sig. (2-tailed)
Pair 1	100 Freestyle 1 - 100 Freestyle 2	5.9931	7.0858	1.2526	3.4384	8.5478	4.785	31	.000

We find the means for each event, the correlation between the two variables, and the test results for the mean difference between races. On average, swimmers improved their times by 5.9931 seconds; the test statistic (t) is +4.785, and the *P*-value is approximately 0. We interpret the test statistic and *P*-value just as we did in the one-sample case.

What do you conclude? Was the observed difference statistically significant? How do you decide?
As noted earlier, we correctly treat this as a paired sample *t* test. But what would happen if we were to (mistakenly) treat the data as two *independent* samples? Doing so requires rearranging the data file considerably, so we'll just look at the consequences of setting this up as an independent samples test. Here are the results:

T-Test

Group Statistics

	First or second heat	N	Mean	Std. Deviation	Std. Error Mean
100 Freestyle time	1	32	76.1072	10.1949	1.8022
	2	32	70.1141	9.0255	1.5955

Independent Samples Test

		Levene's Test for Equality of Variances		t-test for Equality of Means						95% Confidence Interval of the Difference	
		F	Sig.	t	df	Sig. (2-tailed)	Mean Difference	Std. Error Difference	Lower	Upper
100 Freestyle time	Equal variances assumed	.147	.708	2.490	62	.015	5.9931	2.4070	1.1816	10.8046
	Equal variances not assumed			2.490	61.102	.016	5.9931	2.4070	1.1802	10.8060

How does this result compare to prior one? In the correct version of this test, we saw a large, significant increase—the *t* statistic was nearly 5. **Why is this result so different?**
Under slightly different circumstances, the difference between the two test results could be profound. This example illustrates the importance of knowing which test applies in a particular case. Any software package is able to perform the computations either way, but the onus is on the analyst to know which method is the appropriate one. As you can see, getting it "right" makes a difference—a large effect nearly "disappears" when viewed through the lens of an inappropriate test.

Moving On...

Use the techniques presented in this lab to answer the following research questions. Justify your conclusions by citing appropriate test statistics and *P*-values. Explain your choice of test procedure. Unless otherwise noted, use $\alpha = .05$.

Student

Before conducting these tests, write a brief prediction of what you expect to see when you do the test. Explain why you might or might not expect to find significant differences for each problem.

1. Do commuters and residents earn significantly different mean grades?

2. Do car owners have significantly fewer accidents, on average, than nonowners?

3. Do dog owners have fewer siblings than nonowners?

4. Many students have part-time jobs while in school. Is there a significant difference in the mean number of hours of work for males and females who have such jobs? (Omit students who do not have outside hours of work.)

Colleges

5. For each of the variables listed below, *explain why you might expect to find significant differences between means for public and private colleges.* Then, test to see if there is a significant difference between public and private colleges.
   - Mean combined SAT scores of incoming freshmen
   - Percent of incoming freshmen in the top 10% of their high school class
   - Number of full-time undergraduate students
   - Tuition charges for in-state students
   - Percent of students who graduate within four years

Swimmer2

6. Do individual swimmers significantly improve their performance between the first and second recorded times in the 50-meter freestyle?

7. In their second times, do swimmers who compete in the 50-meter freestyle swim faster than they do in the 50-meter backstroke?

Water

8. Is there statistically significant evidence here that water resources subregions were able to reduce irrigation conveyance losses (i.e., leaks) between 1985 and 1990?

9. Did mean per capita water use change significantly between 1985 and 1990?

World90

10. Two of the variables in this file (rgdpch and rgdp88) measure real Gross Domestic Product (GDP) per capita in 1990 and 1988. Is there statistically significant evidence that per capita GDP increased between 1988 and 1990?

11. On average, do these 42 countries tend to invest more of their GDP than they devote to government spending?

GSS94

12. Is there a statistically significant difference in the amount of time men and women spend watching TV?

13. Is there a statistically significant difference in the amount of time married and unmarried people spend watching TV? (You'll need to create a new variable to represent the two groups here.)

BP

14. Do subjects with a parental history of hypertension have significantly higher resting systolic blood pressure than subjects with no parental history?

15. Do subjects with a parental history of hypertension have significantly higher resting diastolic blood pressure than subjects with no parental history?

Infant

This dataset comes from research conducted by Dr. Lincoln Craton, who has been investigating cognition during infancy. In this study, he was interested in understanding infants' ability to perceive stationary, partially hidden objects. He presented infants of various ages the scenes depicted in the figure below.

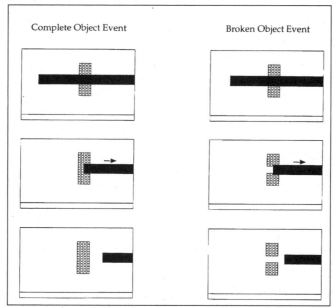

When shown the top left panel, most adults report seeing a vertically positioned rectangle behind a long horizontal strip. What do you suppose a 5- or 8-month-old infant sees here?

Infants in this study were presented with the two test events shown above. The primary dependent variable in this study was looking-time. Prior research has shown that infants tend to look longer at the more surprising of two events. Do infants find the broken event to be more surprising, as adults do, who have a strong tendency to perceive partially hidden objects as complete?

16. Do 5-month-old infants perceive a partially hidden object as complete? First, select infants who are between 5 and 6 months old (defined here as between 500 and 530 in age; 500 is read as 5 months and 00 days). Compare these infants' total looking-time at the broken event (totbroke) to their total looking-time at the complete event (totcompl). Is there a significant difference? What does this result mean?

17. Do the same analysis with infants who are about 8 months old (defined in this study between 700 and 830 in age). Do 8-

month-old infants perceive a partially hidden object as complete? In other words, do they look differently at the broken event (totbroke) than they do at the complete event (totcompl)? What does this result mean? How does this result compare to that found in the 5-month-old infants?

Physique

This file contains data collected by two research method students who were interested in investigating the effects of social physique anxiety, that is, a person's interpretation of how another perceives one's physique. More specifically, they wondered whether females' social physique anxiety would impact how comfortable they felt in various social situations. Social physique anxiety (SPA) was measured using a scale that contained statements such as, "I am comfortable with how fit my body appears to others." Situational comfort levels were measured using a scale that contained items such as "When I am giving an oral presentation I feel comfortable." Both scales used the following statement responses: not at all, slightly, moderately, very, extremely.

18. Do women who score high on the SPA show significantly more discomfort in social situations (total) than women who score low on the SPA? NOTE: SPA scores were used to form the two groups (spalevel), where high represents scores above the median and low represents scores below the median.

London2

This file contains carbon monoxide (CO) measurements in West London air by hour for all of 1996. Measurements are parts per million.

19. Would you expect to find higher CO concentrations at 9 AM or at 5 PM? Perform an appropriate t test and comment on what you find. How do you explain the result? What might account for different levels of carbon monoxide during different times in the day?

20. Would you expect to find higher CO concentrations at 9 AM or at 9 PM? Perform an appropriate t test and comment on what you find. How do you explain the result? What might account for different levels of carbon monoxide during different times in the day?

Analysis of Variance (I)

Objectives

In this session, you will learn to do the following:
- Perform and interpret a one-factor independent measures analysis of variance (ANOVA)
- Understand the assumptions necessary for a one-factor independent measures ANOVA
- Perform and interpret post-hoc tests for a one-factor independent measures ANOVA
- Perform and interpret a one-factor repeated measures ANOVA
- Understand the assumptions necessary for a one-factor repeated measures ANOVA

Comparing Three or More Means

In Session 12, we learned to perform tests that compare one mean to another. When we compared samples drawn from two independent populations, we performed an *independent-samples t test*. When just one sample of subjects was used to compare two different treatment conditions, we performed a *paired-samples t test*.

It is often the case, however, that we want to compare three or more means to one another. Analysis of variance (ANOVA) is the procedure that allows for such comparisons. Like the *t* tests previously discussed, ANOVA procedures are available for both independent measures tests and repeated measures tests. There are a number of ANOVA procedures, typically distinguished by the number of *factors*, or independent variables, involved. Thus, a one-factor ANOVA (sometimes called one-way ANOVA) indicates there is a single independent variable. We begin with the independent measures ANOVA where we are comparing three or more samples from independent populations.

One-Factor Independent Measures ANOVA

Suppose we were curious about whether students who have to work many hours outside of school to support themselves find their grades suffering. We could examine this question by comparing the GPAs of students who work various amounts of time outside of school (e.g., many hours, some hours, no hours). The one factor in this example is the amount of work because it defines the different conditions being compared.

Let's examine this question using the data in our **Student** file. Open that data file now. One variable is called WorkCat and represents work time outside of school (0 hours, 1–19 hours, 20 or more hours). For a first look at the average GPA for each of the three work categories, do the following:

🖱 **Graphs ➤ Boxplot...** In the Boxplot dialog box, choose Simple and click **Define**. In the next dialog box, select GPA as the Variable, and WorkCat for the Category Axis. Click on **Options...**, click off Display groups defined by missing values, and click **Continue**. Then click **OK**.

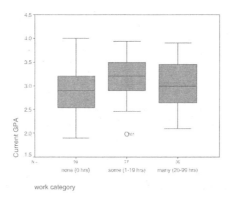

The boxplots show some variation across the groups, with the highest GPAs belonging to the students who worked between 1 and 19 hours. As you look in the Viewer window, you should notice that the median GPA (the dark line in the middle of the box) differs slightly among the groups, but then again, the sample medians of any three groups of students would differ to some degree. That's what we mean by sampling error.

Thus the inferential question is, should we attribute the observed differences to sampling error, or do they reflect genuine differences among the three populations? Neither the boxplots nor the sample medians offer decisive evidence.

Part of the reason we perform a formal statistical test like ANOVA is to clarify some of the ambiguity. The ANOVA procedure will distinguish how much of the variation we can ascribe to sampling error and how much to the *factor* (the demands of a part-time job, in this case).

In this instance, we would initially hypothesize no difference among the mean GPAs of the three groups. Formally, our null and alternative hypotheses would be:

H_o: $\mu_1 = \mu_2 = \mu_3$
H_A: at least one population mean is different from the others

Before performing the analysis, however, we should review the assumptions required for this type of ANOVA. Independent measures ANOVA requires three conditions for reliable results:

- Independent samples
- Normal populations
- Homogeneity (or equality) of population variances

We will be able to formally test both the normality and the homogeneity-of-variance assumptions. With large samples, homogeneity of variance is more critical than normality, but one should test both. We can examine the normality assumption both graphically and by the use of a formal statistical test.

Analyze ➤ Descriptive Statistics ➤ Explore... In the Explore dialog box, select GPA as the Dependent List variable, WorkCat as the Factor List variable, and Plots as the Display. Next, click on **Plots...**

The **Explore** command offers a number of choices. We want to restrict the graphs to those most helpful in assessing normality. Several default options will be marked in the Explore Plots dialog box. Since we are interested in a normality test only, just select Normality plots with tests, as shown on the next page:

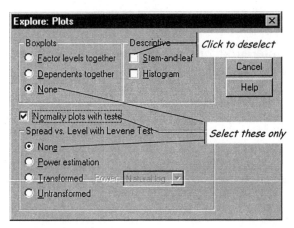

Click on **Continue** and then click **OK** in the main dialog box.

The output from this set of commands consists of several parts.[1] We will focus our attention on the tests of normality as shown here:

Tests of Normality

	work category	Kolmogorov-Smirnov[a]			Shapiro-Wilk		
		Statistic	df	Sig.	Statistic	df	Sig.
Current GPA	none (0 hrs)	.073	99	.200*			
	some (1-19 hrs)	.068	77	.200*			
	many (20-99 hrs)	.100	36	.200*	.960	36	.338

*. This is a lower bound of the true significance.

a. Lilliefors Significance Correction

This test is computed only when n < 50

The Kolmogorov-Smirnov test assesses whether there is a significant departure from normality in the population distribution for each of the three groups. The null hypothesis states that the population distribution is normal. Look at the test statistic and significance columns for each of the three work categories. The test statistics range from .068 to .100 and the *P*-values (significance) are all .200. Since these *P*-values are greater than our α (.05), we do not reject the null hypothesis and conclude that these data do not violate the normality assumption.

We still need to validate the homogeneity-of-variance assumption. We do this within the ANOVA command. Thus, we may proceed with our one-factor independent measures ANOVA as follows:

---

[1] The normal and detrended plots represent additional ways to assess normality in the form of graphs.

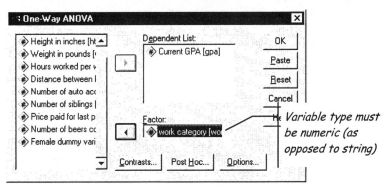

Analyze ➤ Compare Means ➤ One-Way ANOVA... The Dependent List variable is GPA and the Factor variable is WorkCat.

Click on **Options...** and under Statistics, select Descriptive, so we can look at the group means, and Homogeneity-of-variance. Click on **Continue** and then click on **OK** in the main dialog box.

The One-Way ANOVA output consists of several parts. Let's first look at the test for homogeneity of variances, since satisfying this assumption is necessary for interpreting ANOVA results.

Test of Homogeneity of Variances

Current GPA

Levene Statistic	df1	df2	Sig.
.414	2	209	.662

Levene's test for homogeneity of variances assesses whether the population variances for the groups are significantly different from each other. The null hypothesis states that the population variances are equal. The Levene statistic has a value of .414 and a P-value of .662. We interpret the P-value as we did before with the normality test: since P is greater than α (.05), we do not reject the null hypothesis and conclude that these data do not violate the homogeneity-of-variance assumption. Later, in Session 21, we'll see what we do when the assumption is violated.

Having concluded that we have indeed met the assumptions of the independent measures ANOVA, let's find out whether students who work various amounts of time outside of school differ in their GPAs. To do so, look at the ANOVA table at the bottom of your output (it should look very similar to the ANOVA summary tables shown in your textbook).

ANOVA

Current GPA

	Sum of Squares	df	Mean Square	F	Sig.
Between Groups	3.558	2	1.779	8.865	.000
Within Groups	41.943	209	.201		
Total	45.501	211			

You can see that the test statistic (F) equals 8.865 with a corresponding P-value of .000. In this test, we would reject the null hypothesis, and conclude that these data provide substantial evidence of a least one significant difference in mean GPAs among the three groups of students.

Where Are the Differences?

After performing a one-factor independent measures ANOVA and finding out that the results are significant, we know that the means are not all the same. This relatively simple conclusion, however, actually raises more questions: Is μ_1 different than μ_2? Is μ_1 different than μ_3? Is μ_2 different than μ_3? Are all three means different? Post-hoc tests provide answers to these questions whenever we have a significant ANOVA result.

There are many different kinds of post-hoc tests, that examine which means are different from each other. One commonly used procedure is Tukey's Honestly Significant Difference test. The Tukey test compares all pairs of group means without increasing the risk of making a Type I error.[2]

Before performing a Tukey test using our data, let's look at the group means to get an idea of what the GPAs of the various work categories look like. The default ANOVA procedure does not display the group means. To get them, we needed to select them as an option, as we did above. Our output contains the group means in the Descriptives section.

---

[2] Typically, when we perform a series of hypothesis tests, we increase the risk of a Type I error beyond the α we have set for one test. For example, for one test, the risk of a Type I error might be set at α = .05. However, if we did a series of three tests, the risk of a Type I error would increase to α = .15 (3 x .05), which is usually deemed too high.

Descriptives

Current GPA

	N	Mean	Std. Deviation	Std. Error	95% Confidence Interval for Mean		Minimum	Maximum
					Lower Bound	Upper Bound		
none (0 hrs)	99	2.8800	.4584	4.608E-02	2.7886	2.9714	1.90	4.00
some (1-19 hrs)	77	3.1664	.4270	4.866E-02	3.0695	3.2633	1.98	3.94
many (20-99 hrs)	36	3.0197	.4626	7.710E-02	2.8632	3.1763	2.10	3.90
Total	212	3.0077	.4644	3.189E-02	2.9449	3.0706	1.90	4.00

Which group had the lowest mean GPA? Which group had the highest mean GPA? Do you think a mean GPA of 3.17 is significantly better than a mean GPA of 3.02?

As we have learned, eyeballing group means cannot tell us decisively if significant differences exist. We need statistical tests to draw definitive conclusions. Let's run the Tukey test on our data to find out where the differences are. To do so, we return to the One-Way ANOVA dialog box:

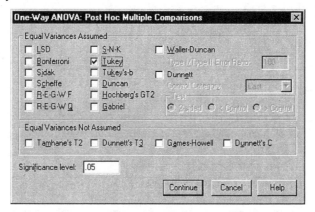 **Analyze ➤ Compare Means ➤ One-Way ANOVA...** The variables are still selected, as earlier. Click on **Post-Hoc...,** and select only Tukey, as shown here:

Your output will contain all the parts of the original one-way ANOVA analysis that we just discussed, as well as several new tables. We will focus our attention on the multiple comparisons table as shown on the next page.

The first line of the table represents the pairwise comparison of the mean GPAs between the none and the some categories of work. The mean difference is listed as –.2864 and an asterisk (*) is displayed next to it, indicating that this represents a significant difference. Looking back at the group means in the Descriptives section, we can conclude that students who worked some hours (1–19 hrs.) had better GPAs than students who did not work at all. ***Does this result surprise you? Why?***

Multiple Comparisons

Dependent Variable: Current GPA
Tukey HSD

(I) work category	(J) work category	Mean Difference (I-J)	Std. Error	Sig.	95% Confidence Interval	
					Lower Bound	Upper Bound
none (0 hrs)	some (1-19 hrs)	-.2864*	6.807E-02	.000	-.4459	-.1268
	many (20-99 hrs)	-.1397	8.719E-02	.245	-.3441	6.462E-02
some (1-19 hrs)	none (0 hrs)	.2864*	6.807E-02	.000	.1268	.4459
	many (20-99 hrs)	.1466	9.045E-02	.237	-6.53E-02	.3586
many (20-99 hrs)	none (0 hrs)	.1397	8.719E-02	.245	-6.46E-02	.3441
	some (1-19 hrs)	-.1466	9.045E-02	.237	-.3586	6.534E-02

*. The mean difference is significant at the .05 level.

Try interpreting the rest of the table in this manner. Note that each mean is compared to every other mean twice (e.g., $\mu_1-\mu_2$ and $\mu_2-\mu_1$) so the results are essentially repeated in the table. The results of this Tukey test could be summarized as follows:

(1) Students who worked some hours (1–19 hrs.) had better GPAs than students who did not work at all
(2) Students who worked some hours (1–19 hrs.) had comparable GPAs to students who worked many hours (20 + hrs.)
(3) Students who worked many hours had comparable GPAs to students who did not work at all

How would you explain these results? Can you think of another variable that might be a contributing factor in these results?

One-Factor Repeated Measures ANOVA

> ⌨ The repeated measures ANOVA procedure is not part of the SPSS Base (rather it is part of the SPSS Advanced Models option). It is possible that your system will not be able to run this procedure. Consult your instructor about the availability of the SPSS Advanced Models option on your system.

The prior example focused on comparing three or more samples from independent populations. There are many situations, however, where we are interested in examining the same sample across three or more treatment conditions. These tests are called *repeated measures analysis of variance* because several measurements are taken on the same set of individuals.

For example, suppose we were interested, as Dr. Christopher France was, in whether blood pressure changes during various stressor tasks. We could examine this question by comparing blood pressure measurements in a sample of individuals across several types of tasks. The one factor in this example is the type of stressor because it defines the different conditions being compared.

Let's examine this question using the data in a file called **BP.** Open this data file now. This file contains data about blood pressure and other vital signs during various physical and mental stressors. These data were collected in a study investigating factors associated with the risk of developing high blood pressure, or hypertension. The subjects were all college students.

We will perform a test to see if there is evidence that diastolic[3] blood pressure changes significantly during three different conditions: resting, doing mental arithmetic, and immersing a hand in cold water. Our null hypothesis would state that blood pressure does not change during the stressors. Formally, our null and alternative hypotheses would be:

H_0: $\mu_1 = \mu_2 = \mu_3$
H_A: at least one population mean is different from the others

As before, we should review the conditions required for this type of ANOVA. Repeated measures ANOVA requires four conditions for reliable results:

- Independent observations within each treatment
- Normal populations within each treatment
- Equal population variances within each treatment
- Sphericity (discussed later)

We will formally assess, with a statistical test, both the normality assumption and the sphericity assumption. In general, researchers are

---

[3] Blood pressure is measured as blood flows from and to the heart. It is reported as a systolic value and a diastolic value. If your pressure is 130 over 70, the diastolic pressure is 70.

not overly concerned with the assumption of normality except when small samples are used.

Because these are repeated measures performed on each student, the dependent variable (blood pressure) is represented by three variables:

- Dbprest: diastolic blood pressure at rest
- Dbpma: diastolic blood pressure during a mental arithmetic task
- Dbpcp: diastolic blood pressure while immersing a hand in ice water

🖰 **Analyze ➤ Descriptive Statistics ➤ Explore...** In the Explore dialog box, select dbprest, dbpma, and dbpcp as the Dependent List variables and Plots as the Display. Next, click on **Plots...**

🖰 Select None under Boxplots, select Normality plots with tests, and deselect Stem-and-leaf under Descriptive. Click on **Continue** and then click **OK** in the main dialog box.

Tests of Normality

	Kolmogorov-Smirnov[a]		
	Statistic	df	Sig.
DBPREST	.046	175	.200*
DBPMA	.036	175	.200*
DBPCP	.056	175	.200*

*. This is a lower bound of the true significance.

a. Lilliefors Significance Correction

We interpret the Kolmogorov-Smirnov tests for normality just as we did earlier. ***What are the results of these normality tests? Can we feel confident that we have not violated the normality assumption? How do you decide?***

Although this is a one-factor test, the fact that it uses repeated measures means we use a different ANOVA command.

🖰 **Analyze ➤ General Linear Model ➤ Repeated Measures...** The dialog box shown on the next page will appear, prompting us to assign a name to our repeated measure (also called within-subject factor) and indicate the number of conditions (also called levels) we have.

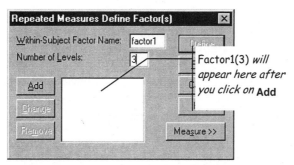

SPSS uses the default name factor1 for the Within-Subjects Factor Name and we will leave it as such.

🖰 Type in 3 as the Number of Levels and click on **Add**. Factor1(3) will appear in the adjacent box. Now click on **Define** and another dialog box will appear prompting us to select the specific conditions of our repeated measures factor, called factor1.

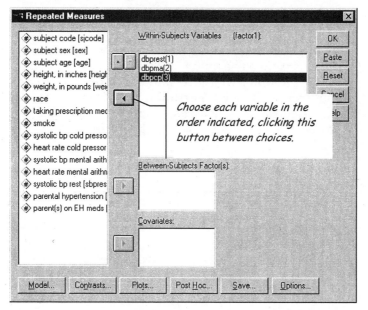

🖰 Select dbprest as the first level of factor1 and dbpma and dbpcp as the second and third levels, respectively. Click on **Options...** and select Descriptive Statistics (to get the treatment means). Click on **Continue** and then click **OK** in the main dialog box.

The Repeated Measure ANOVA output consists of several parts. Let's first look at Mauchly's test of sphericity, since the results of it will determine which type of ANOVA test should be used. Essentially, the null hypothesis in the sphericity assumption is that the correlations among the three diastolic blood pressure measures be equal. Let's look at the results of the sphericity test.

Mauchly's Test of Sphericity[b]

Measure: MEASURE_1

Within Subjects Effect	Mauchly's W	Approx. Chi-Square	df	Sig.	Epsilon[a] Greenhouse-Geisser	Huynh-Feldt	Lower-bound
FACTOR1	.654	73.415	2	.000	.743	.748	.500

Tests the null hypothesis that the error covariance matrix of the orthonormalized transformed dependent variables is proportional to an identity matrix.

a. May be used to adjust the degrees of freedom for the averaged tests of significance. Corrected tests are displayed in the Tests of Within-Subjects Effects table.

b.
Design: Intercept
Within Subjects Design: FACTOR1

The test statistic for Mauchly's test of sphericity is the chi-square statistic[4] and in this case has a value of 73.415 and a P-value of .000. We therefore reject the null hypothesis and conclude that we have *not* met the assumption of sphericity. Fortunately, SPSS provides several alternative tests when the sphericity assumption has been violated in a repeated measures ANOVA. Take a look at the output labeled Tests of Within-Subjects Effects.

Tests of Within-Subjects Effects

Measure: MEASURE_1

Source		Type III Sum of Squares	df	Mean Square	F	Sig.
FACTOR1	Sphericity Assumed	9467.806	2	4733.903	73.754	.000
	Greenhouse-Geisser	9467.806	1.486	6370.949	73.754	.000
	Huynh-Feldt	9467.806	1.496	6329.115	73.754	.000
	Lower-bound	9467.806	1.000	9467.806	73.754	.000
Error(FACTOR1)	Sphericity Assumed	22336.396	348	64.185		
	Greenhouse-Geisser	22336.396	258.580	86.381		
	Huynh-Feldt	22336.396	260.289	85.814		
	Lower-bound	22336.396	174.000	128.370		

---

[4] Chi-square tests are presented in Session 20.

This table has the columns of an ANOVA summary table (e.g., Sum of Squares, df, Mean Square, F, Sig.) with several additional rows. The first line of factor1 reads "Sphericity Assumed." This would be the F test line you would interpret if the sphericity assumption has been met. Since this was not the case for us, we look at one of the other lines to determine whether diastolic blood pressure changes significantly during the various mental and physical stressors investigated in this study. The second, third, and fourth lines of factor1 represent several kinds of F tests where adjustments have been made because sphericity has been violated.[5] The Greenhouse-Geisser adjusted F test is commonly used so we will interpret this one here.

You can see that the test statistic (*F*) equals 73.754, with a corresponding *P*-value of .000. ***What is your conclusion based on this test? Does diastolic blood pressure change significantly during the various mental and physical stressors investigated in this study?***

Where Are the Differences?

Recall that at the conclusion of a significant ANOVA, whether it be independent measures or repeated measures, we know that the means of the groups or conditions are not equal. But we don't know where the differences are. Post-hoc tests are designed to examine these differences.

Unfortunately, for our purposes, SPSS does not offer post-hoc tests for repeated measures ANOVA. This is likely due to the fact that statisticians don't agree what the error term should be for these tests. Consult your instructor on how to proceed when you encounter a significant one-factor repeated measures ANOVA. He or she may suggest performing a Tukey test by hand (using MSerror in the place of MSwithin) or doing a series of paired-samples *t* tests.

Moving On...

Use the techniques you have learned in this session to answer these questions. Determine what kind of one-factor ANOVA (independent measures or repeated measures) is appropriate. Then check the

---

[5] The adjustments are made in the degrees of freedom that are used to evaluate the significance of the *F* statistic. Note the slight differences in df in the various columns. Also notice the fact that the same conclusion regarding significance would be drawn from any of these F tests (the original or any of the adjusted tests).

underlying assumptions for that procedure. Explain what you conclude from your full analysis and why. Use $\alpha = .05$.

GSS94

This is the extract from the 1994 General Social Survey.

1. One variable in the file groups respondents into one of four age categories. Does the mean number of television hours vary by age group?

2. Does the amount of television viewing vary by a respondent's subjectively identified social class?

MFT

This dataset contains Major Field Test (MFT) results, SAT scores and GPAs for college seniors majoring in a science. Department faculty are interested in predicting a senior's MFT performance based on high school or college performance. The variables GPAQ, VerbQ, and MathQ indicate the quartile in which a student's GPA, verbal SAT, and math SAT scores fall within the sample.

3. Do mean total scores on the MFT vary by GPA quartile? Comment on distinctive features of this ANOVA.

4. Does the relationship between total score and GPA hold true for each individual portion of the MFT?

5. Do mean total scores on the MFT vary by verbal SAT quartile? math SAT quartile?

6. Based on the observed relationships between GPA and MFT, one faculty member suggests that college grading policies need revision. Why might one think that, and what do you think of the suggestion?

Milgram

This file contains the data from several famous obedience studies by Dr. Stanley Milgram. He wanted to better understand why people so often obey authority, as so many people did during the Holocaust.

Each experiment involved an authority figure (Dr. Milgram), a teacher (a male subject), and a learner (a male accomplice of the experimenter). Dr. Milgram told the teacher that he would be asking the learner questions while the learner would be connected to a shock generator under the teacher's control. Dr. Milgram instructed the teacher to shock the learner for a wrong answer, increasing the level of shock for each successive wrong answer. The shock generator was marked from 15 volts (slight shock) to 450 volts (severe shock); the teacher received a sample 45-volt shock at the outset of the experiment. The main measure for any subject was the maximum shock he administered.

Milgram performed several variations of this experiment. In experiment 1, the teacher could not hear the voice of the learner because the two were in separate rooms. In experiment 2, the teacher and the learner were in separate rooms but the teacher could hear the learner's yells and screams. In experiment 3, the teacher and the learner were in the same room, only a few feet from one another. In experiment 4, the teacher had to hold the learner's hand on the shock plate.

Please note: the learner never received any shocks, but the teacher was not at all aware of this. In fact, the situation caused the teachers so much duress that this kind of experiment would not be permissible today under the ethical guidelines established by the American Psychological Association.

7. Although these experiments were run as separate studies, could we look at them as four conditions in an independent measures ANOVA? That is, could we examine the question whether proximity to the learner had an effect on the maximum amount of shock delivered by the teacher? Would we be wise to do so? Start by checking the assumptions.

BP

This dataset contains data from individuals whose blood pressure and other vital signs were measured while they were performing several physical and mental stressor tasks.

8. Does systolic blood pressure change significantly during the three tasks examined in this study? The three tasks were: at rest (sbprest), performing mental arithmetic (sbpma), and immersing a hand in ice water (sbpcp).

Anxiety2

This file contains data from a study examining whether the anxiety a person experiences affects performance on a learning task. Subjects with varying levels of anxiety performed a learning task across a series of four trials and the number of errors made was recorded.

9. Regardless of anxiety level, does the number of errors made by subjects change significantly across the learning trials (trial1, trial2, trial3, trial4)?

Nielsen

10. Does the mean Nielsen rating vary by television network?

AIDS

11. Did the 1992 AIDS rates vary significantly by WHO region? Did the 1993 rates do so?

Airline

This is data collected from major airlines throughout the world. It contains information on crash rates and general geographic regions for each airline.

12. Is there a difference among the geographic regions in crash rates per million flight miles? Comment on what you find and offer some explanations for your conclusions about airlines from different geographic regions.

Analysis of Variance (II)

Objectives

In this session, you will learn to do the following:

- Perform and interpret a two-factor independent measures analysis of variance (ANOVA)
- Understand the assumptions necessary for a two-factor independent measures ANOVA
- Understand and interpret statistical main effects
- Understand and interpret statistical interactions

Two-Factor Independent Measures ANOVA

In Session 13, we learned how to perform analysis-of-variance procedures for research situations where there is a single independent variable. There are many situations, however, where we want to consider *two independent variables at the same time*. The analysis for these situations is called two-factor ANOVA. Like the one-factor procedures, there are two-factor ANOVA procedures for both repeated measures designs and independent measures designs. Our focus here will be on two-factor independent measures ANOVA, as the repeated measures variety is beyond the scope of this book.

Let's consider a research study by Dr. Christopher France, who was interested in risk factors for developing hypertension. Prior research had found that people at risk showed changes in cardiovascular responses to various stressors. Dr. France wanted to explore this finding further by looking at several variables that might be implicated, in particular, a person's sex and whether the person had a parent with hypertension. These two independent variables resulted in four groups of participants: males with parental hypertension, males without parental hypertension, females with parental hypertension, and females without parental hypertension.

There were various dependent variables measured, but we will focus on systolic blood pressure during a mental arithmetic task. In particular, subjects were asked to add or subtract two- and three-digit numbers that were presented for five seconds.

The analysis appropriate for these data is a two-factor independent measures ANOVA. It is a *two-factor* ANOVA because there are two independent variables (sex and parental history); it is an *independent measures* ANOVA because the samples come from independent populations.

The analysis of a two-factor ANOVA actually involves three distinct hypothesis tests. Specifically, the two-factor ANOVA will test for:

(1) The mean difference between levels of the first factor (in our data, this would be comparing systolic blood pressure during mental arithmetic [sbpma] for males and females).

(2) The mean difference between levels of the second factor (in our data, this would be comparing sbpma for individuals with parental hypertension and individuals without parental hypertension).

(3) Any other mean differences that may result from the unique combination of the two factors (in our data, one sex might show distinct systolic blood pressure only when they have a parent who has hypertension).

The first two hypothesis tests are called tests for the *main effects*. The null hypothesis for main effects is always that there are no differences among the levels of the factor (e.g., H_0: $\mu_{males} = \mu_{females}$). The third hypothesis test is called the test for the *interaction*, because it examines the effects of the combination of the two factors together. The null hypothesis for the interaction is always that there is no interaction between the factors (e.g., H_0: the effect of sex does not depend on parental history and the effect of parental history does not depend on sex).

It should be noted that these three hypothesis tests are independent tests. That is, the outcome of one test does not impact the outcome of any other test. Thus, it is possible to have any combination of significant and nonsignificant main effects and interactions.

Let's explore a two-factor independent measures ANOVA now, bearing in mind that the assumptions for this test are the same as those required for the one-factor independent measures ANOVA (i.e., independent samples, normal populations, and equal population variances).

🖱 Open the file called **BP**.

🖱 **Analyze ➤ General Linear Model ➤ Univariate...** Select systolic bp mental arithmetic [sbpma] as the Dependent Variable and sex and parental hypertension [PH] as the Fixed Factors.

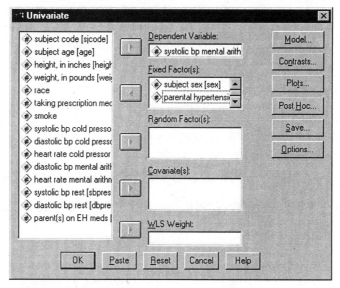

🖱 As we did in other ANOVA analyses, click on **Options...** and under Display, select Descriptive statistics and Homogeneity tests. Click on **Continue** and then click on **OK** in the main dialog box.

The Univariate ANOVA output consists of several parts (descriptive statistics, Levene's test of equality of variances, and the tests of between-subjects effects). Let's first look at the Levene's test (see next page) for homogeneity of variances, since satisfying this assumption is necessary for interpreting ANOVA results in any meaningful way.

Levene's Test of Equality of Error Variances

Dependent Variable: systolic bp mental arithmetic

F	df1	df2	Sig.
1.364	3	174	.255

Tests the null hypothesis that the error variance of
the dependent variable is equal across groups.

a. Design: Intercept+SEX+PH+SEX * PH

Levene's test for homogeneity of variances assesses whether the population variances for the groups are significantly different from each other. The Levene statistic (F) has a value of 1.364 and a P-value of .255. Since P is greater than α (.05), we do not reject the null hypothesis and we conclude that these data do not violate the homogeneity-of-variance assumption. If this assumption had been violated, we would not proceed with interpretation of the ANOVA.

Now let's find out whether systolic blood pressure during a mental arithmetic task is related to a person's sex, parental history of hypertension, or some combination of these factors. To do so, look at the table at the bottom of your output. This table has the columns of an ANOVA summary table with several additional rows.

Tests of Between-Subjects Effects

Dependent Variable: systolic bp mental arithmetic

Source	Type III Sum of Squares	df	Mean Square	F	Sig.
Corrected Model	7286.555[a]	3	2428.852	15.541	.000
Intercept	2917864.694	1	2917864.694	18669.573	.000
SEX	5704.665	1	5704.665	36.501	.000
PH	1439.365	1	1439.365	9.210	.003
SEX * PH	215.986	1	215.986	1.382	.241
Error	27194.434	174	156.290		
Total	2949638.444	178			
Corrected Total	34480.989	177			

a. R Squared = .211 (Adjusted R Squared = .198)

Remember that a two-factor ANOVA consists of three separate hypothesis tests, the results of which are listed in this table. Locate the line labeled SEX and notice that the F statistic for this test of the main effect has a value of 36.501, with a corresponding significance of .000.

We reject the null hypothesis and conclude that there is a significant main effect for the SEX factor.

Let's continue reading the ANOVA table before we try to make sense of the results. Locate the line labeled PH. ***Is there a significant main effect for the Parental History factor? How do you know?***

To find out if there is a significant interaction between the Sex and Parental History factors, we read the line labeled SEX * PH (usually read as SEX by PH). Notice that the *F*-value is 1.382, with a *P*-value of .241. Thus, we do not reject the null hypothesis and conclude there is no significant interaction between the two factors.

What exactly do these results mean? We have two significant main effects and a nonsignificant interaction. One very helpful way to make sense of two-factor ANOVA results is to graph the data.

Graphs ➤ Interactive ➤ Bar... First drag Count (SPSS default) back to the variable list. Then drag sbpma to the vertical axis, PH to the horizontal axis, and sex to the Panel Variables.

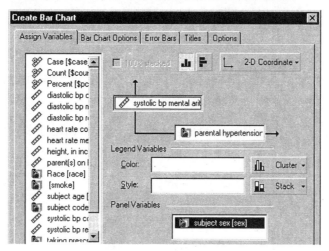

Click on the Titles tab to title your bar chart and put your name in the caption.

Look at your bar chart and notice that all the bars are about the same height, so the significant main effects are hard to discern. Let's use the interactive graphing capabilities to make a bar chart where the *y* axis (sbpma) does not start with zero, and thus shows the differences more clearly. According to the Descriptives section of the ANOVA output, the minimum sbpma is 118; let's start the *y* axis at 115.

🖰 Double-click anywhere on your bar chart. Find the Chart Manager icon at the top of your bar chart and click on it. You will see the familiar Chart Manager dialog box.

🖰 In the Chart Manager, click on **Scale Axis** and then on **Edit...**

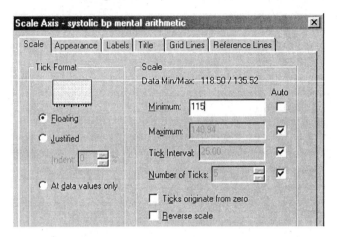

🖰 In the Scale Axis dialog box, under Scale, find Minimum and deselect Auto. Type in 115. Click **OK**.

🖰 Close the Chart Manager dialog box by clicking the ☒ button in the upper right of the dialog box.

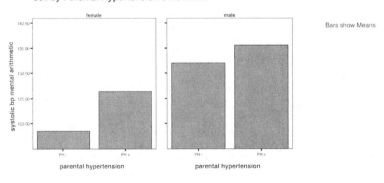

Prepared by R. Carver & J. Nash

This new bar chart shows the significant main effects much more clearly. You can see in your bar chart that males have higher systolic blood pressure during mental arithmetic than females, regardless of parental hypertension. You can also see that individuals having a parent with hypertension have higher systolic pressure than those who do not, and this occurs regardless of gender. Thus, we can conclude that both factors separately, but not their combination, can put one at risk for developing hypertension.

Another Example

Let's explore another two-factor independent measures ANOVA in our attempt to more thoroughly understand main effects and interactions. This next set of data comes from a study conducted by Dr. Bonnie Klentz, who was interested in understanding the dynamics of individuals working in small groups. More specifically, she wanted to investigate whether people can accurately perceive fellow group members' output in a task. Two variables she thought might be important were the size of the work group and the gender of the participant. These two independent variables made up the four groups in the study: all-male groups of 2, all-male groups of 3, all-female groups of 2, and all-female groups of 3.

Participants in this study were asked to find as many words as they could in a matrix of 16 letters (for any Boggle® fans out there, the task was modeled on that game). Each time they found a word, participants wrote it on a slip of paper and put it in a box in the center of the table. Participants were told that the total number of words found by each group would be counted at the end of the study. Unbeknownst to the participants, the paper sizes of each group member differed so that the number of slips put in by each individual could also be computed by the researcher.

We will focus here on participants' estimation of their coworkers output (i.e., how many words do you think your coworkers found?). The accuracy of this estimation was calculated as a difference score—the actual number of slips of paper put in the box minus the estimated number of slips of paper put in the box. A positive difference score indicates that participants were underestimating their coworker's performance.

Let's find out whether gender, group size, or a combination of these two factors impacts people's accuracy of estimating their coworker's performance. To do so, open the file called **Group**.

✒ **Analyze ➤ General Linear Model ➤ Univariate...** Select difnext as the Dependent Variable and gender and grpsize as the Fixed Factors. As we did before, click on **Options...** and under Display, select Descriptive statistics and Homogeneity tests. Click on **Continue** and then click on **OK** in the main dialog box.

Let's first assess whether the homogeneity-of-variance assumption has been met. Levene's test shows the F statistic has a value of 2.571, with a corresponding P-value of .061. ***Have we violated the assumption of homogeneity of variance? How do you decide?***

To find out if either factor or a combination of the two impacts perception of a coworker's output, we need to evaluate the ANOVA summary table.

Tests of Between-Subjects Effects

Dependent Variable: subject's perception of co-worker

Source	Type III Sum of Squares	df	Mean Square	F	Sig.
Corrected Model	608.646[a]	3	202.882	3.572	.018
Intercept	261.750	1	261.750	4.609	.035
GENDER	83.519	1	83.519	1.471	.229
GRPSIZE	24.554	1	24.554	.432	.513
GENDER * GRPSIZE	256.569	1	256.569	4.518	.037
Error	4032.101	71	56.790		
Total	5240.000	75			
Corrected Total	4640.747	74			

a. R Squared = .131 (Adjusted R Squared = .094)

Is there a main effect for gender? Is there a main effect for group size? How do you know? Locate the line for the interaction of Gender by Group Size and notice that the F statistic has a value of 4.518, with a corresponding P-value of .037. Thus, we will reject the null hypothesis for this test and conclude there is a significant interaction between gender and group size.

To illustrate these results, we will graph our data as we did before.

✒ **Graphs ➤ Interactive ➤ Bar...** First, drag Count back to the variable list. Then drag difnext to the vertical axis, grpsize to the horizontal axis, and gender to the Panel Variables. Be sure to put an appropriate title and caption on your bar chart.

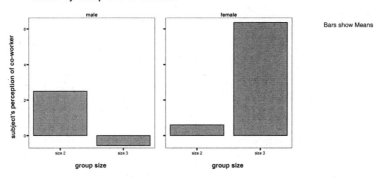

Prepared by R. Carver & J. Nash

This bar chart nicely displays the interaction between the gender and group size factors. Recall that the dependent variable is the difference between the actual performance of the coworkers and the participants' estimation of their work. Thus, a positive number indicates an underestimation of the coworker's performance. Notice that males underestimated their coworker's performance when working in groups of two whereas females underestimated their coworker's performance when working in groups of three. Thus, group size did influence the ability to estimate coworker's performance but the effect was *different* for males and females. This is exactly what is meant by a statistical interaction—the effects of one factor depend on the effects of the other factor.

Do these results make sense to you? How would you explain these findings? It is interesting to note that in this original study, the researcher only predicted a main effect for group size. Specifically, she predicted that participants in groups of two would be more accurate in estimating their partner's productivity than participants in groups of three. This was based on the idea that in larger groups, there is more information to attend to. As we know from our analysis above, this hypothesized main effect was not supported, and instead a surprising interaction appeared. And that's what keeps researchers collecting and analyzing data, because the answer to one question often raises another question to explore.

One Last Note

As it turned out in the examples in this session, results showed either the presence of significant main effects (the **BP** data) or the presence of a significant interaction (the **Group** data), but not the combination of both a significant interaction and significant main effect(s). Because each of the F tests in a two-factor ANOVA is independent, this combination is certainly possible. When such a situation arises, it is important to begin interpretation with the significant interaction because an interaction can distort the main effects. For this reason, statisticians say, "interactions supersede main effects."

Moving On...

Use the techniques of this session to respond to these questions. Check the underlying assumptions for the two-factor ANOVA. Also, explain what you conclude from the analysis, using graphs to help illustrate your conclusions. Use $\alpha = .05$.

BP

Recall that this dataset contains blood pressure and other vital signs during various physical and mental stressors.

1. Is heart rate while immersing a hand in ice water (hrcp) related to a person's sex, parental hypertension (PH), or some combination of these factors?

2. Is heart rate while performing mental arithmetic (hrma) related to these same factors (sex, parental hypertension, or their combination)?

3. What happens to diastolic blood pressure during mental arithmetic (dbpma) as it relates to a person's sex, parental hypertension, or their combination? How do these results compare to those found in this session's example that used systolic blood pressure during mental arithmetic (sbpma) in a 2 (SEX) by 2 (PH) ANOVA?

Group

Recall that this file contains data from a study investigating the dynamics of individuals working in small groups.

4. Is the actual productivity of subjects (subtot1) related to the size of group they are working in (grpsize), their gender, or a combination of these factors? In other words, which of these factors or their combination makes a difference in how hard someone works in a small group?

Haircut

This dataset comes from the **Student** data, which was collected on the first day of class. Students were asked the last price they paid for a professional haircut. In addition, they were asked to specify the region where they got that haircut, according to the following categories: rural, suburban, or urban.

5. Is the price of a haircut related to a person's sex, the region where they got the haircut, or some combination of these factors?

Census90

This is a sample of the 1990 Census, including various demographic and personal information.

6. One might suspect that level of education and gender both have significant impacts on one's salary. Is this true? Comment on what you find. Is there anything confusing about what you find? NOTE: Select only those cases for which income is greater than 0. Be sure to check assumptions.

Student

Recall that these data are collected from first day business students and contain demographic and personal information.

7. Propose a theory to explain why both gender and major field might affect one's GPA. Using this set of data, test your theory.

8. Does gender and one's rating of personal driving ability affect the number of accidents one has been in during the past year? Comment on noteworthy features of this analysis.

Linear Regression (I)

Objectives

In this session, you will learn to do the following:
- Perform a simple, two-variable linear regression analysis
- Evaluate the goodness of fit of a linear regression model
- Test hypotheses concerning the relationship between two quantitative variables

Linear Relationships

Some of the most interesting questions of statistical analysis revolve around the relationships between two variables. How many more traffic fatalities will occur in a state as more cars share the highways? How much will regional water consumption increase if the population increases by 1,000 people?

In each of these examples, there are two common elements—a *pair of quantitative variables* (e.g., water consumption and population), and a *theoretical reason* to expect that the two variables are related.

Linear regression analysis is a tool with several important applications. First, it is a way of *testing hypotheses* concerning the relationship between two numerical variables. Second, it is a way of *estimating* the specific nature of such a relationship. Beyond asking, "Are water consumption and population related?" regression allows us to ask *how* they are related. Third, it allows us to *predict* values of one variable if we know or can estimate the other variable.

As a first illustration, consider the classic economic relationship between consumption and income. Each additional dollar of income enables a person to spend (consume) more. As income increases, we expect consumption to rise as well. Let's begin by looking at aggregate income and consumption of all individuals in the United States, over a long period of time.

🖱 Open the **US** data file. This file contains different economic and demographic variables for the years 1965–1996. We'll examine Aggregate personal consumption [perscon] and Aggregate personal income [persinc], which represent the total spending and total income, respectively, for everyone in the United States.

If we were to hypothesize a relationship between income and consumption, it would be positive: the more we earn as a nation, the more we can spend. Formally, the theoretical model of the relationship might look like:

Consumption = Intercept + (slope)(Income) + random error,   or

$$y = \beta_0 + \beta_1 x + \varepsilon, \quad \text{where } \varepsilon \text{ is a random error term.}[1]$$

If x and y genuinely have a positive linear relationship, β_1 is a positive number. If they have a negative relationship, β_1 is a negative number. If they have *no relationship at all,* β_1 is zero.

First, let's construct a scatterplot of the two variables. Our theory says that consumption depends on income. In the language of regression analysis, consumption is the *dependent* variable and income is the *independent* variable. It is customary to plot the dependent variable on the y axis, and the independent on the x axis.

🖱 **Graphs ➤ Interactive ➤ Scatterplot...** The y axis variable is Aggregate personal consumption [perscon] and the x axis variable is Aggregate personal income [persinc]. Click **OK**.

As you look at the resulting plot (facing page), you can see that the points fall into nearly a perfect straight line. This is an example of pronounced *positive* or *direct* relationship, and a good illustration of what a linear relationship looks like. It is called a positive relationship because the line has a positive, or upward, slope. One interpretation of the phrase "linear relationship" is simply that x and y form a line when graphed. But what does that mean in real-world terms? It means that y changes by a *constant amount* every time x increases by one unit.

---

[1] The random error term, ε, and the assumptions we make about it are discussed more fully in Session 16.

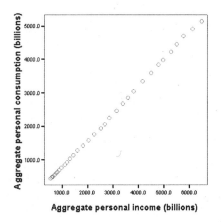

In this graph, the points form a nearly a perfect line. The regression procedure will estimate the equation of that line which comes closest to describing the pattern formed by the points.

Analyze ➤ Regression ➤ Linear... The Dependent variable is consumption, and the Independent is income.

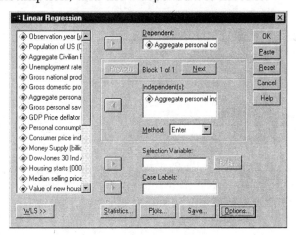

Look in your Viewer window. The regression output consists of four parts: a table of variables in the regression equation, a model summary, an ANOVA table, and a table of coefficients. Your text may deal with some or all of these parts in detail; in this discussion, we'll consider them one at a time, focusing on portions of the output. In due time, we'll explain all of the results.

Variables Entered/Removed[b]

Model	Variables Entered	Variables Removed	Method
1	Aggregate personal income (billions)[a]	.	Enter

a. All requested variables entered.

b. Dependent Variable: Aggregate personal consumption (billions)

We will think of a regression equation as a model that explains or predicts variation in a dependent variable. The table of Variables Entered/Removed lists the independent variable in the model. As we will see later, it is possible to have several independent variables, and we may want to examine regression models that contain different combinations of those variables. Thus, SPSS refers to variables having been "entered" into or "removed" from a model, and anticipates the possibility that there are several models within a single analysis. In this example, there is just one variable: aggregate personal consumption.

We know that the regression procedure, via the *least squares method* of estimation[2], gives us the line that fits the points better than any other. We might ask just how "good" that fit is. It may well be the case that the "best fitting" line is not especially close to the points at all!

Model Summary

Model	R	R Square	Adjusted R Square	Std. Error of the Estimate
1	1.000[a]	.999	.999	35.316

a. Predictors: (Constant), Aggregate personal income (billions)

The second standard part of the regression output—the Model Summary—reports a statistic that measures "goodness of fit." The statistic is called the *coefficient of determination*, represented by the symbol r^2. It is the square of r, the coefficient of correlation, which is also reported here. Locate R Square on your screen. For now, ignore the Adjusted R Square; it is used in multiple regression, and is discussed in Session 17.

---

[2] Consult your primary text for an explanation of the least squares method.

r^2 can range from 0.000 to 1.000, and indicates the extent to which the line fits the points; 1.000 is a perfect fit, such that each point is on the line. The higher the value of r^2, the better. In this example, we can see that changes in income account for 99.9% of the variation in consumption.

ANOVA[b]

Model		Sum of Squares	df	Mean Square	F	Sig.
1	Regression	7.1E+07	1	7.1E+07	56929.002	.000[a]
	Residual	37415.665	30	1247.189		
	Total	7.1E+07	31			

a. Predictors: (Constant), Aggregate personal income (billions)

b. Dependent Variable: Aggregate personal consumption (billions)

The next element in the output is an ANOVA table. You should recognize this from earlier sessions, and should remember that certain assumptions must be satisfied before interpreting ANOVA results. The assumptions associated with regression analysis are treated fully in Session 16. At this point in the discussion, we will focus on how the table is used when the assumptions are satisfied.

Recall we said earlier that, if x and y were unrelated, the slope of the best-fitting line would be 0. When we run the regression procedure, we compute an *estimated* slope. Typically, this slope is nonzero. It is critical to recognize that the estimated slope is a result of the particular sample at hand. A different sample would yield a different slope. Thus, our estimated slope is subject to sampling error, and therefore, is a matter for hypothesis testing.

In this instance, the null hypothesis being tested is that the true slope, β_1, equals 0. Here, with an F statistic in excess of 56,000 and a significance level of 0, we would reject the null.

Coefficients[a]

Model		Unstandardized Coefficients		Standardized Coefficients	t	Sig.
		B	Std. Error	Beta		
1	(Constant)	-38.586	11.244		-3.432	.002
	Aggregate personal income (billions)	.809	.003	1.000	238.598	.000

a. Dependent Variable: Aggregate personal consumption (billions)

The final piece of output is the table of coefficients. In a regression equation, the slope and the intercept are referred to as the *coefficients* in the model. We find both coefficients in the column labeled

B; the intercept (β_0) is the constant, and the slope (β_1) is the coefficient of the aggregate personal income. From the table, we find that our estimated line can be written as:

$$\text{Consumption} = -38.586 + .809\text{Income}$$

The slope of the line (.809) means that if Personal Income increases by one billion dollars, Personal Consumption increases by 0.809 billion dollars. In other words, in the aggregate, Americans consumed about 81 cents of each additional dollar earned.[3]

What does the intercept mean? The value of –38.586 literally says that if income were 0, consumption would be –38.586 billion dollars. This makes little sense, but reflects the fact our dataset lies very far from the y axis. In essence, the estimated y-intercept is a huge extrapolation beyond the observed data. The line which we estimate must cross the axis somewhere; in this instance, it crosses at –38.586.

The table of coefficients also reports some t-statistics and significance levels. We'll return to those in our next example.

Another Example

Traffic fatalities are a tragic part of American life. They occur all too often across the United States, but with varying frequency from state to state. In this example, we will begin to develop a model to account for the varying number of fatalities among the states hypothesizing that states with large populations will tend to have more traffic fatalities than sparsely populated states. Open the data file called **States**.

🖱 **Graphs ➤ Interactive ➤ Scatterplot...** This time, Auto accident fatalities [accfat] is y, and Population, 1994 [pop] is x.

Compare this graph to the scatterplot of consumption and income. How would you describe this relationship?

Clearly, the connection between fatalities and population is not as strong as the connection between consumption and income. We'll run a regression analysis to evaluate the relationship.

🖱 **Analyze ➤ Regression ➤ Linear...** Select the fatalities variable as the dependent, and the population variable as the independent.

---

[3] If you've studied economics, you may recognize this as the *marginal propensity to consume*. Note that the slope refers to the *marginal* change in y, given a one-unit change in x. It is *not* true that we spent 81 cents of *every* dollar. We spend 81 cents of the next dollar we have.

The model summary indicates that the correlation between Auto Accident Fatalities and Population is 0.950. **What does this correlation coefficient tell you?**

In the table of coefficients, the reported slope is 1.619E-04, or .0001619. The estimated line is:

Auto Accident Fatalities = 76.190 + .0001619 Population

If one state has 100,000 more residents than a neighboring state, how many more fatalities does this equation estimate in that state? What does the slope of the line tell you about fatalities and population?

Statistical Inferences in Linear Regression

One of the standard tests we perform in a regression analysis is designed to judge whether there is any *significant linear relationship* between x and y. Our null hypothesis is that there is none; formally, the hypotheses look like this:

$$H_0: \beta_1 = 0$$
$$H_A: \beta_1 \neq 0$$

Earlier, we saw that the ANOVA table provides us with the test results. In a regression with one independent variable, we have two equivalent tests—the F and t tests. In the table of coefficients for this regression (below), the right-most columns are labeled t and Sig. These represent t tests asking if the intercept and slope are equal to zero. In this case we cannot reject the null hypothesis for the constant (intercept), because $P \leq .186$. That makes some sense: If a state had a population of 0, we might expect 0 fatalities.

Coefficients[a]

Model		Unstandardized Coefficients		Standardized Coefficients	t	Sig.
		B	Std. Error	Beta		
1	(Constant)	76.190	56.773		1.342	.186
	Population, 1994	1.619E-04	.000	.950	20.989	.000

a. Dependent Variable: Auto Accident Fatalities

The value of the test statistic for the slope is 20.989, and the associated P-value is approximately 0.[4] As in all t tests, we take this to

---

[4] Very observant students will note that $t^2 = F$.

mean that we should *reject* our null hypothesis, meaning *there is a statistically significant relationship* between *x* and *y*.

So, we have a line that fits the points rather well, and evidence of a statistically significant relationship. To better visualize how this line fits the points, do the following:

🖱 **Graphs ➤ Interactive ➤ Scatterplot...** You have already created a scatterplot for these variables, but now we want to add in the least squares line. In the dialog box, click on the Fit tab; in the drop-down list under Method, select Regression, and click **OK**.

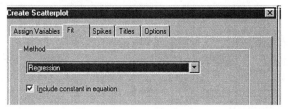

The resulting graph includes the best-fitting least squares line with the scatterplot. According to this regression, differences in the population of states account for approximately 90% of the variation in the number of automobile accident fatalities.

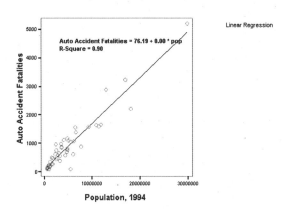

An Example of a Questionable Relationship

We've just seen two illustrations of fairly strong, statistically significant linear relationships. Let's look at another example, where the theory is not as compelling.

Nutrition and session experts concern themselves with a person's body fat, or more specifically the percentage of total body weight made up of fat. Measuring body fat precisely is more complicated than measuring other body characteristics, so it would be nice to have a mathematical model relating body fat to some easily measured human attribute. Our dataset called **Bodyfat** contains precise body fat and other measurements of a random sample of adult men. Open the data file.

Suppose we wondered if height could be used to predict the percentage of body fat. We could use these data to investigate that relationship.

    Graphs ➤ Interactive ➤ Scatterplot... *Y* is Percent % body fat [fatperc] and *x* is Height in inches [height]. Note that the regression line appears on this plot; it will appear in all interactive scatterplots until we deselect Regression in the Fit tab. ***Is there evidence in this graph of a positive linear relationship?***

    Analyze ➤ Regression ➤ Linear... The dependent variable is body fat, and the independent is height.

Do the regression results suggest a relationship? What specific parts of the output tell you what you want to know about the connection between body fat percentage and a man's height? Why do you think the regression analysis turned out this way? What, if anything, does the intercept tell you in this regression?

The key point here is that even though we can estimate a least squares line for any pair of variables, we may often find that there is no statistical evidence of a relationship. Neither the scatterplot nor the estimated slope are sufficient to determine significance; we must consult a *t*- or *F*-ratio for a definitive conclusion.

An Estimation Application

The Carver family uses natural gas for heating the house and water, as well as for cooking. Each month, the gas company sends a bill for the gas used. The bill is a treasure trove of information, including two variables which are the focus of this example.

The first is a figure which is approximately equal to the number of cubic feet of natural gas consumed per day during the billing period. More precisely, it equals the number of *therms* per day; a therm is a measure of gas consumption reflecting the fact that the heating capacity of natural gas fluctuates during the year. In the data file, this variable is called Mean therms consumed per day [gaspday].

The second variable is simply the mean temperature for the period (called Mean temperature in Boston [meantemp]). These two variables are contained in the data file called **Utility**. Open that data file now.

We start by thinking about *why* gas consumption and outdoor temperature should be related. **Which variable is the dependent variable? What would a graph of the relationship look like? Do we expect it to be linear? Do we expect the slope to be positive or negative? Before proceeding, sketch the graph you expect to see.**

> ᐧᵇ **Graphs ➤ Interactive ➤ Scatterplot...** Construct a scatterplot with gas consumption on the vertical axis, and temperature on the horizontal. **Does there appear to be a relationship?**

> ᐧᵇ **Analyze ➤ Regression ➤ Linear...** This time, you select the variables appropriately.

Now look at the regression results. **What do the slope and intercept tell you about the estimated relationship? What does the negative slope indicate? Is the estimated relationship statistically significant? How would you characterize the goodness of fit?**

One fairly obvious use for a model such as this is to predict or estimate how much gas we'll use in a given month. For instance, in a month averaging temperatures of 40 degrees, the daily usage could be computed as:

$$gaspday = 15.368 - 0.217 \, (40)$$
$$= 15.368 - 8.68$$
$$= 6.688 \text{ therms per day.}$$

Use the model to estimate daily gas usage in a month when temperatures average 75°. **Does your estimate make sense to you? Why does the model give this result?**

A Classic Example

Between 1595 and 1606 at the University of Padua, Galileo Galilei (1564–1642) conducted a series of famous experiments on the behavior of projectiles. Among these experiments were observations of a ball rolling down an inclined ramp (see diagram on next page). Galileo varied the height at which the ball was released down the ramp, and then measured the horizontal distance which the ball traveled.

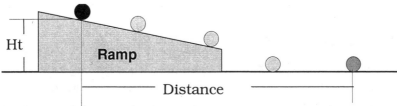

We'll begin by looking at the results of one of Galileo's experiments; the data are in **Galileo**. As you might expect, balls released at greater heights traveled longer distances. Galileo hoped to discover the relationship between release height and horizontal distance. Both the heights and distances are recorded in *punti* (points), a unit of distance.

   First, make a scatterplot of the data in the first two columns of the worksheet, with horizontal distance as the *y* variable.

Does the graph suggest that distance and height are related? Is the relationship positive or negative? For what physical reasons might we expect a nonlinear relationship?
Although the points in the graph don't quite fall in a straight line, let's perform a linear regression analysis for now.

   Perform the regression, using horizontal distance as the dependent variable.

Using the regression results, comment on the meaning and statistical significance of the slope and intercept, as well as the goodness-of-fit measures. Use the estimated regression equation to determine the release height at which a ball would travel 520 punti.
Look back at your scatterplot. *Do you think your estimated release height for a 520 punti travel is probably high or low? Explain.*
It should be clear that a linear model is not the very best choice for this set of data. Regression analysis is a very powerful technique, which is easily misapplied. In upcoming sessions, we'll see how we can refine our uses of regression analysis to deal with problems such as nonlinearity, and to avoid abuses of the technique.

Moving On...

Use the techniques and information in this session to answer the following questions. Explain or justify your conclusions with appropriate graphs or regression results.

Galileo

1. Galileo repeated the rolling ball experiment with slightly different apparatus, described in Appendix A. Use the data in the third and fourth columns of the worksheet to estimate the relationship between horizontal distance and release height.

2. At what release height would a ball travel 520 *punti* in this case?

US

Investigate possible linear relationships between the following pairs of variables. *In each case*, comment on (a) why the variables might be related at all, (b) why the relationships might be linear, (c) the interpretation of the estimated slope and intercept, (d) the statistical significance of the model estimates, and (e) the goodness of fit of the model. (In each pair, the y variable is listed first.)

3. Aggregate Personal Savings vs. Aggregate Personal Income

4. Cars in use vs. Population

5. Total Federal Receipts vs. Aggregate Personal Income

6. GDP vs. Aggregate Civilian Employment

States

7. In the session, we used the Population variable to predict the number of Auto accident fatalities. Run a new regression analysis with Number of registered vehicles as the independent variable, and compare the results of this regression to your earlier model.

8. Do the same using the Number of licensed drivers as the independent variable.

MFT

These are the Major Field Test scores, with student GPA and SAT results. Investigate possible linear relationships between the following pairs of variables. *In each case*, comment on (a) why the variables might be related at all, (b) why the relationships might be linear, (c) the interpretation of the estimated slope and intercept, (d) the statistical significance of the model estimates, and (e) the goodness of fit of the model. (In each question, the y variable is the total MFT score.)

9. GPA

10. Verbal SAT

11. Math SAT

Bodyfat

12. These are the body fat and other measurements of a sample of men. Our goal is to find a body measurement which can be used reliably to estimate body fat percentage. For each of the three measurements listed here, perform a regression analysis. Explain specifically what the *slope* of the estimated line means in the context of body fat percentage and the variable in question. Select the variable which you think is best to estimate body fat percentage.
    - Chest circumference
    - Abdomen circumference
    - Weight

13. Consider a man whose chest measurement is 95 cm, abdomen is 85 cm, and who weighs 158 pounds. Use your best regression equation to estimate this man's body fat percentage.

Track

This file contains the NCAA best-recorded indoor and outdoor times in the women's 3,000-meter track event for 1999.

14. Run an analysis to see if we can use a woman's best indoor time to predict her best outdoor time. Report the results of your analysis.

15. Discuss some possible reasons for the rather poor goodness-of-fit statistics.

Impeach

This file contains the results of the Senate impeachment trial of President Clinton. Each senator could have cast 0, 1, or 2 guilty votes in the trial.

16. The file contains a rating by the American Conservative Union for each senator. A very conservative senator would have a rating of 100. Run a regression using number of guilty votes cast as the dependent variable, and the conservatism rating as the independent. Based on this result, would you say that political ideology was a good predictor of a senator's vote?

17. The file also includes the percentage of the vote received by President Clinton in the senator's home state during the 1996 election. Run a regression to predict guilty votes based on this variable; based on this result, would you say that electoral politics was a good predictor of a senator's vote?

18. Comment on the appearance of the scatterplots appropriate to the prior two questions. Does linear regression seem to be an ideal technique for analyzing these data? Explain.

Water

This file contains freshwater usage data for 221 regions throughout the United States in 1985 and 1990.

19. Construct a regression model that uses 1985 total freshwater withdrawals to estimate the 1990 total freshwater withdrawals. Comment on the meaning and possible usefulness of this model for water planners.

20. Construct a regression model that uses 1985 domestic consumptive use (i.e. household water) to estimate the 1990 total freshwater withdrawals. Comment on the meaning and possible usefulness of this model for water planners. Compare these results to those in the prior question. Which model is better? Why might it have turned out that way?

Bowling

These are the results of one night's bowling for a local league. Each bowler rolls a "series" made up of three separate "strings." The series total is just the sum of the three strings for each person.

21. Develop a simple regression model that uses a bowler's first string score to estimate his or her series total for the evening. Why might we (or might we not) expect a linear relationship between a first-string score and the series total?

Linear Regression (II)

Objectives

In this session, you will learn to do the following:
- Interpret the standard error of the estimate (*s*)
- Validate the assumptions for least squares regression by analyzing the *residuals* in a regression analysis
- Use an estimated regression line to estimate or predict *y* values

Assumptions for Least Squares Regression

In the prior session, we learned to fit a line to a set of points. SPSS uses a common technique, called the *method of least squares*.[1] Though there are several other alternative methods available, least squares estimation is by far the most commonly used.

We can apply the technique to any set of paired (*x*, *y*) values and get an estimated line. However, if we plan to use our estimates for consequential decisions, we want to be sure that the estimates are unbiased and otherwise reliable. The least squares method will yield unbiased, consistent, and efficient[2] estimates when certain conditions

---

[1] This method goes by several common names, but the term *least squares* always appears, referring to the criterion of minimizing the sum of squared deviations between the estimated line and the observed *y* values.

[2] You may recall the terms *unbiased, consistent,* and *efficient* from earlier in your course. This is a good time to review these definitions.

are true. Recall that the basic linear regression model states that x and y have a linear relationship, but that any observed (x, y) pair will randomly deviate from the line. Algebraically, we can express this as:

$$y_i = \beta_0 + \beta_1 x_i + \varepsilon_i$$

where

> x_i, y_i represent the ith observation of x and y, respectively,
> β_0 is the intercept of the underlying linear relationship,
> β_1 is the slope of the underlying linear relationship, and
> ε_i is the ith random disturbance (i.e., the deviation between the theoretical line and the observed value [x_i, y_i])

For least squares estimation to yield reliable estimates of β_0 and β_1, the following must be true about ε, the random disturbance.[3]

- *Normality*: At each possible value of x, the random disturbances are normally distributed; $\varepsilon | x_i$ follows a normal distribution.
- *Zero mean*: At each possible value of x, the mean of $\varepsilon | x_i$ is 0.
- *Homogeneity of variance* (also called *homoskedasticity*): At each possible value of x, the variance of $\varepsilon | x_i$ equals σ^2, which is constant.
- *Independence*: At each possible value of x, the value of $\varepsilon_i | x_i$ is independent of all other $\varepsilon_j | x_j$.

If these conditions are not satisfied and we use the least squares method, we run the risk that our inferences—the tests of significance and any confidence intervals we develop—will be misleading. Therefore, it is important to verify that we can assume that x and y have a linear relationship, and that the four above conditions hold true. The difficulty lies in the fact that we cannot directly observe the random disturbances, ε_i, since we don't know the location of the "true" regression line. In lieu of the disturbances, we instead examine the *residuals*—the differences between our estimated regression line and the observed y values.

Examining Residuals to Check Assumptions

By computing and examining the residuals, we can get some idea of the degree to which the above conditions apply in a given regression analysis. We will adopt slightly different analysis strategies depending on

---

[3] Some authors express these as assumptions concerning $y | x_i$.

whether the sample data are cross-sectional or time series. A *cross-sectional* sample is drawn from a population at one point in time; *time series*, or *longitudinal*, data involves repeated measurement of a sample across time.

In cross-sectional data, the assumption of independence is not relevant since the observations are not made in any meaningful sequence; in time series data, though, the independence assumption is important. We will start with a cross-sectional example that we saw in the prior session: the relationship between traffic fatalities and the population of a state. Open the data file called **States**. This time when we perform the regression, we'll have SPSS report some additional statistics, and generate graphs to help us evaluate the residuals.

    Create a scatterplot, including the fitted line, with fatalities on the vertical axis and population on the horizontal.

    Analyze ➤ Regression ➤ Linear... As you did before, select auto accident fatalities as the dependent, and 1994 population as the independent variable. Before clicking **OK**, click on **Statistics....** This button opens another dialog box, allowing you to specify various values for SPSS to report. Complete the dialog box as shown here, and then click **Continue**.

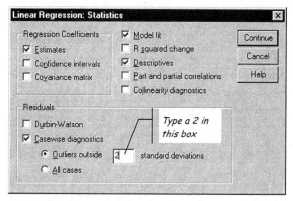

In this dialog box, we ask for two outputs beyond the defaults. First, we will generate descriptive statistics for the two variables in the regression. These will include mean, standard deviation, and correlations. Second, we will generate a table of those cases whose residuals are more than two standard deviations from the estimated line. This can help us to identify outlying values in the regression.

In addition to this information, we will want to evaluate the normality and homoskedasticity assumptions by creating two graphs of our residuals. We do so in the following way:

Next, click on **Plots...** and complete the dialog box as shown here.

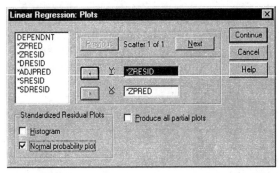

This dialog box allows us to create numerous graphs. As shown, it will generate two graphs: a normal probability plot for assessing the normality assumption and a scatterplot of standardized predicted values (*ZPRED) versus standardized residuals (*ZRESID). We will explain the interpretation of the graphs soon.

We interpret the regression results exactly as in the prior session. As you may recall, this regression model looks quite good: the significance tests are impressive, and the coefficient of determination (r^2) is quite high. Before examining the residuals *per se*, we call your attention to several elements of the regression output.

Descriptive Statistics

	Mean	Std. Deviation	N
Auto Accident Fatalities	889.56	926.34	50
Population, 1994	5025067	5434552	50

Shown above are the descriptive statistics for the two variables in the regression. These are helpful in thinking about the relative magnitude of the residuals, and about some of the model summary measures. In particular, locate the Model Summary on your screen. The right-most value in the table is called the Standard Error of the Estimate (sometimes called *s*), and equals 293.38 fatalities. This value is a measure of the variability of the random disturbance. Assuming normality, about 95% of the time the observed points should fall within two standard errors of the regression line. If s were close in size to the standard deviation of y, we would encounter as much estimation error predicting y with our model as with simply using the mean value of y. Since s is so much smaller than the standard deviation of y (926 in this example), we can feel confident that our model is an improvement over the naïve estimation of auto fatalities using the mean value of y.

Look further down in the regression output and find the table labeled Casewise Diagnostics. Three cases are listed here because their standardized residuals[4] exceed 2 in absolute value. The first has a very large negative residual. This state had only 95 fatalities, but the regression equation would have predicted over 966. **Can you identify this point in the scatterplot that you made earlier? Can you surmise the identity of the state? What are the other two states (cases 10 and 33)?**

Casewise Diagnostics[a]

Case Number	Std. Residual	Auto Accident Fatalities	Predicted Value	Residual
2	-2.971	95	966.50	-871.50
10	2.460	2892	2170.36	721.64
33	-2.646	2212	2988.17	-776.17

a. Dependent Variable: Auto Accident Fatalities

Recall that there are four assumptions about the random errors in the regression equation. We cannot observe the random errors themselves, but instead we will examine the residuals to help evaluate whether any of the four least squares assumptions should be questioned. We can inspect the residuals as surrogates for the random disturbances, and make judgments about three of the four assumptions.

The least squares method guarantees that the *residuals* will always have a mean of zero. Therefore, the mean of the residuals carries no information about the mean of the random errors.

In this regression, we need not concern ourselves with the fourth (independence) assumption since we are dealing with cross-sectional data, as noted earlier. However, we will examine the residuals to help decide if the other two assumptions—*normality* and *homogeneity of variance*—are valid. The simplest tools for doing so are the two graphs we requested. At this point, we will do no formal testing for these properties, but simply provide some guidelines for interpreting the graphs.

---

[4] A standardized residual is simply the residual expressed as a *z*-score.

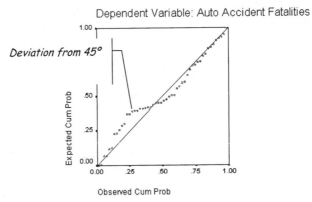

Normal P-P Plot of Regression Standardized Residual

Dependent Variable: Auto Accident Fatalities

The first graph is called a Normal Probability Plot. If the residuals are normally distributed, they will lie along the 45° upward sloping diagonal line. Our null hypothesis is that the residuals *are* normal, and to the extent that the graphs deviate substantially from the 45° pattern, the normality assumption should be questioned. In this case, the lower left portion of the graph raises some concerns.

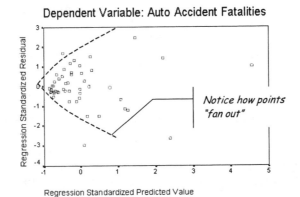

Scatterplot

Dependent Variable: Auto Accident Fatalities

The next graph is a scatterplot of the standardized residuals versus the standardized fitted, or estimated, values[5]. This graph can give us insight into the assumption of equal variances, as well as the

---

[5] Some authors prefer to plot residuals versus *x* values; the graphs are equivalent.

assumption that x and y have a linear relationship. When both are true, the residuals will be randomly scattered in an even, horizontal band around a residual value of zero, as illustrated here in this idealized plot.

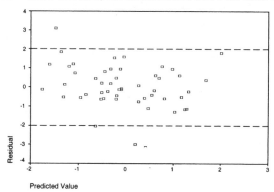

Residuals that "fan out" from left to right, or right to left signal *heterogeneity of variance* (or *heteroskedasticity*). A curved pattern suggests a nonlinear relationship. In our regression, we see residuals that are increasingly dispersed as we read across from left to right. This suggests that the variances are not constant along the regression line, and this is bad news for us. As with other techniques, a violation of the homogeneity-of-variance assumption (homoskedasticity) renders our inferences unreliable. We should not interpret the significance tests in this particular regression. Sessions 17 and 18 suggest some strategies for dealing with heteroskedasticity.

A Time Series Example

Earlier, we noted that the assumption of independence is often a concern in time series datasets. If the disturbance at time $t + 1$ depends on the disturbance at time t, inference and estimation present some special challenges. Once again, our initial inclination will be to assume that the random disturbances *are* independent, and look for compelling evidence that they are not. As before, we do so by running a least squares regression, saving the fitted and residual values, and examining the residuals.

Our next example uses time series data, so that the sequence of these observations is meaningful. As in the prior session, we'll look at some annual data from the United States, continuing with our automotive theme. It seems plausible to think that, as the U.S.

population has grown, the number of cars on the road has likewise grown. Let's consider that relationship. Open the file **US**.

🖱 **Graphs ➤ Interactive ➤ Scatterplot...** Create a scatterplot with Cars in use (millions) [cars] on the vertical and Population of US (000) [pop] on the horizontal axis. Include a least squares regression line. ***Does this appear to be a linear relationship?***

Notice that there is a series of points below the line, followed by a series above, another below, and so forth. This pattern suggests that the residuals are not independent of one another. As we move through time, positive and negative residuals tend to cluster. When residuals are independent, they fluctuate randomly through time. Repeated "runs" of positive and negative residuals suggest that residuals are dependent on one another.

🖱 **Analyze ➤ Regression ➤ Linear...** The dependent variable is Cars in use (millions) [cars] and the independent variable is Population of US (000) [pop]. As in the prior regression, we want to generate additional statistics and some graphs. Click on **Statistics...**, and complete the dialog box as you did earlier.[6]

🖱 Click on **Save**, and check Standardized under Residuals.

🖱 Now click on **Plots**, and request the same graphs as in the previous example, and run the regression.

🖱 When the regression output has appeared, switch back to the Data Editor, and scroll all the way to the right. You'll see a new variable called zre_1. This is the standardized residuals values.

🖱 **Graphs ➤ Sequence...** Create a sequence, or time series plot of these residuals by completing the dialog box shown on the next page.

---

[6] In this dialog box, you have the option of asking for a Durbin-Watson statistic. The Durbin-Watson statistic can be used to test for the presence of positive serial correlation (i.e., nonindependence), and requires the use of a table that appears in many standard textbooks. Since the table is not universally provided, we will not demonstrate the use of the test here. Your instructor may wish to do so, though.

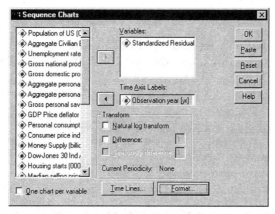

In our graph (shown below), we added a reference line at 0. Notice the pattern of a run of negative residuals, followed by positives, which are in turn followed by negatives. This strongly indicates that the residuals are *not* independent of one another.

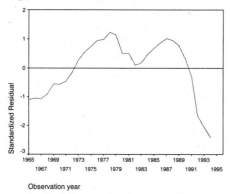

Now look back at the scatterplot and the residual plots. ***What do you conclude about the assumptions of linearity, normality, and homoskedasticity? Should you interpret the significance tests in this regression? Explain.***

Issues in Forecasting and Prediction

One reason that we are concerned about the assumptions is that they affect the reliability of estimates or forecasts. To see how we can use SPSS to make and evaluate such forecasts, we'll turn to another example. Open the file called **Utility**.

This file contains time series data about the consumption of natural gas and electricity in the Carver home. Also included is the mean monthly temperature for each month in the sample. As in the prior session, we'll model the relationship between gas consumption and temperature, and then go on to forecast gas usage. We'll suppose that we want to forecast usage in a month when mean temperature is 22 degrees (as it was in February 1995).

As before, we could just plug 22 degrees into our estimated regression model to get a point estimate. However, we can also develop either a confidence or a prediction interval, depending on whether we wish to estimate the *mean* gas use in all months averaging 22 degrees, or the actual gas use in one particular month averaging 22 degrees.

🖱 **Analyze ➤ Regression ➤ Linear...** As before, Mean therms consumed per day [gaspday] is the dependent variable and Mean Temperature [meantemp] is the independent. As before, we'll ask for the same statistics and plots, but this time, also click on **Save....** Complete the Save dialog box as shown here, and run the regression.

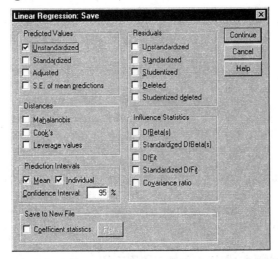

Before examining the regression output, switch to the Data Editor. Scroll down to row 54 (February 1995), and note that the mean temperature was actually 22° and an average of 10.7 therms were consumed per day. Now scroll to the right, and note some oddly named columns in the data file (illustrated on the next page).

54:month		01-FEB-95							
	elecdays	est	hdd	cdd	newroom	pre_1	lmci_1	umci_1	lici_1	uici_1
50	30	1	288	0	0	3.00113	2.62008	3.38219	.35237	5.64989
51	28	0	478	0	0	4.30288	3.97730	4.62845	1.66153	6.94422
52	32	1	814	0	0	7.77419	7.38455	8.16383	5.12418	10.42420
53	29	0	932	0	0	11.46246	10.79902	12.12590	8.75860	14.16632
54	29	1	1016	0	0	10.59463	10.00294	11.18632	7.90748	13.28179
55	31	0	805	0	0	8.42506	7.99522	8.85490	5.76885	11.08128
56	29	1	561	0	0	6.68940	6.35175	7.02705	4.04654	9.33227
57	30	0	258	25	0	4.30288	3.97730	4.62845	1.66153	6.94422

The five new variables corresponded to the predicted values (computed by using the estimated equation and each observed x value), the lower and upper bounds of a 95% confidence interval for the mean of $y|x$, and the lower and upper bounds of the 95% prediction interval for individual values of $y|x$.

For a month with a mean temperature of 22 degrees, the pre_1 column indicates a predicted value of 10.59463 therms, which was quite close to the actual 10.7 consumed that February. The confidence interval for the mean is (10.00294, 11.18632); we can say that we are 95% confident that the mean consumption is in this interval for all months averaging 22 degrees. The 95% prediction interval is (7.90748, 13.28179). We are 95% confident that gas consumption will fall in this interval for one individual month when the mean temperature is 22 degrees. Note that the mean interval is much narrower than the individual interval, reflecting the fact that we can be more confident with a precise estimate of means than we can for individual values.

We could examine the mean and individual intervals for all possible values of x as follows:

Set up an interactive scatterplot of mean therms consumed vs. mean temperature. In the Fit dialog box, locate the area labeled Prediction Lines. Request Mean and Individual intervals.

The resulting graph (shown on the next page) is a bit confusing at first glance. It shows the scatterplot with five lines superimposed. The central straight line, sloping downward is the estimated regression line. On either side of it are a pair of *confidence bands* and another pair of *prediction bands* corresponding to the mean and individual intervals.

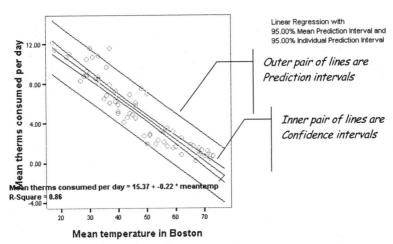

Gas Consumption

Before relying too heavily on these estimates, let's look at the residuals. After all, estimation is a form of inference that depends on the validity of our assumptions. Switch to the Viewer window, and look at the regression results and the residual graphs.

Remember that these are time series data. **For each of the three assumptions we can investigate, do you see any problems? How might these problems affect our predictions in this instance?**

Take another look at the scatterplot of the data and the regression line itself. Along the x axis, visually locate 22 degrees, and look up at the regression line and at the observed data points. Note that most of the observed data from 20 to 30 degrees lies *above* the regression line, and that the full set of points gently arcs around the line. Therefore, estimated values in the low temperature range are probably too low. What can we do about that? We'll see some solutions in a future lab.

A Caveat about "Mindless" Regression

Linear regression is a very powerful tool, which has numerous uses. Like any tool, it must be used with care and thought, though. To see how thoughtless uses can lead to bizarre results, try the following.

   Open the file called **Anscombe.**

This dataset contains eight columns, representing four sets of *x,y* pairs. We want to perform four regressions using x1 and y1, x2 and y2, etc. By now, you should be able to run four separate regressions without detailed instructions. ***After you do so, look closely at the four sets of results, and comment on what you see.***

Based on the regressions, it is tempting to conclude that the four *x-y* pairs all share the same identical relationship. Or is it?

🖱 **Graphs ➤ Interactive ➤ Scatterplot...** Construct four scatterplots (y1 vs. x1, y2 vs. x2, etc.) What do you see? Remember that each of these four plots led to the four virtually identical regressions.

Moving On...

Using the techniques of this session, perform and evaluate regressions and residual analyses to investigate the following relationships. Each file used in this session was also used in Session 15. You may want to refer to those Moving On... questions.

US

Use your regressions and SPSS to predict the dependent variable as specified in the question. In each instance, report the estimated regression equation, and explain the meaning of the slope and intercept.

1. Cars in use vs. Population (predict when pop = 245,000 i.e., 245 million people)

2. Federal Receipts vs. Personal Income (predict when PersInc = 5,000 i.e., 5 trillion dollars)

States

NOTE: The first "unusual" state from the first example is Alaska. In the Data Editor, delete the value in the Pop column for Alaska. Now redo the first example from this session, omitting the unusual case of Alaska.

3. How, specifically, does this affect (a) the regression and (b) the residuals? Compare the slopes and intercepts of your two regressions and comment on what you find.

Bodyfat

4. In Session 15, you did some regressions using these data. This time, perform a linear regression and residual analysis using % Body fat as the dependent variable, and Weight as the independent. Estimate a fitted value when weight = 157 pounds (refer to case #134).

5. Do these sample data suggest that the least squares assumptions have been satisfied in this case?

6. What is the 95% interval estimate of mean body fat percentage among all men who weigh 157 pounds?

7. What is the 95% interval estimate of body fat percentage for one *particular* man who weighs 157 pounds?

Galileo

8. In the previous session, we noted that horizontal distance and release height (first two columns) did not appear to have a linear relationship. Rerun the regression of distance (y) versus height, and construct the residual plots. Where in these plots do you see evidence of nonlinearity?

9. Repeat the same with columns 3 and 4. Is there any problem with linearity here?

MFT

In Session 15, you ran one or more regression models with Total MFT Score as the dependent variable. Repeat those analyses, this time evaluating the residuals for each model.

10. GPA

11. Verbal SAT

12. Math SAT

Track

13. Repeat your regression analysis using indoor time to predict outdoor time. Report on the residuals, noting any possible violations of the least squares assumptions.

Impeach

14. Perform a regression and residual analysis with number of guilty votes as the dependent variable and conservatism rating as the independent variable. Discuss unusual features of these residuals. To what extent do the assumptions seem to be satisfied?

15. Perform a regression and residual analysis with number of guilty votes as the dependent variable and percentage of the vote received by President Clinton in 1996 as the independent variable. Discuss unusual features of these residuals. To what extent do the assumptions seem to be satisfied?

Water

16. Evaluate the validity of the least squares assumptions for a regression model that uses 1985 total freshwater withdrawals to estimate the 1990 total freshwater withdrawals.

17. Evaluate the validity of the least squares assumptions for a regression model that uses 1985 total domestic consumptive use to estimate the 1990 total freshwater withdrawals. How do these residuals compare to those in the previous question?

Bowling

These are the results of one night's bowling for a local league. Each bowler rolls a "series" made up of three separate "strings." The series total is just the sum of the three strings for each person.

18. Develop a simple regression model that uses a bowler's first string score to estimate his or her series total for the evening. Evaluate and comment on the residuals for this regression.

19. Use your regression model to estimate the series score for a bowler who rolls a score of 225 in his first string.

Session 17

Multiple Regression

Objectives

In this session, you will learn to do the following:
- Improve a regression model using multiple regression analysis
- Interpret multiple regression coefficients
- Incorporate qualitative data into a regression model
- Diagnose and deal with multicollinearity

Going Beyond a Single Explanatory Variable

In our previous sessions using simple regression, we examined several bivariate relationships. In some examples, we found a statistically significant relationship between the two variables, but also noted that much of the variation remained *unexplained* by a single independent variable, and that the standard error of the estimate (*s*) was often rather high compared to the standard deviation of the dependent variable.

There are many instances in which we can posit that one variable depends on several others; that is, we have a single effect with multiple causes. The statistical tool of *multiple regression* enables us to identify those variables simultaneously associated with a dependent variable, and to estimate the separate and distinct influence of each variable on the dependent variable.

For example, suppose we want to develop a model to explain the wide variation in college tuition charges. In a simple bivariate model, we might hypothesize that tuition charged by a school depends on the costs incurred by the institution. In our **Colleges** dataset, we have one variable which measures those costs: It is called Instructional expenditure per student [instpers], and represents the per capita expenses directly related to instruction (as opposed to residency, athletics, or other student services) for the school. Let's begin with a simple linear regression of tuition and instructional expenditures.

🖑 **Graphs ➤ Scatter...** Previously we've used the interactive scatterplot command. In this session, as you'll see soon, we want to take advantage of a feature of this command. Define a simple scatterplot with Out-of-state tuition [tuit_out] as the dependent variable, and Instructional expenditure per student [instpers] as the independent.

Comment on the scatterplot, mentioning noteworthy features. Does it seem reasonable to proceed with a linear regression analysis?

🖑 **Analyze ➤ Regression ➤ Linear...** Use tuit_out as the dependent variable, and instpers as the independent. As in earlier sessions, request Descriptives and Casewise diagnostics for residuals more than two standard errors in size, along with a normal probability plot and a plot of standardized residuals vs. standardized predicted values.

Look at the regression results, which are mixed. We find normal residuals, with an oddly-shaped graph of residuals versus predicted values; the F and t test results are therefore suspect, but the reported P-value (Sig. = .000) points to a significant relationship. On the other hand, adjusted r^2 is low at .44 and the standard error of the estimate, $3,126, is fairly large relative to the standard deviation of y. All in all, this model does not fit the data very well.

It appears that instructional costs are associated with about 44% of the variation in out-of-state tuition charges. Another 56% of the variation remains unexplained. Suppose we hypothesize that "better" schools charge more than their peer institutions facing the same instructional costs, other things being equal. Further, let's use a measure of student quality, average combined SAT scores, as our indicator of academic quality. Since we are now interested in the relationship among three variables, a matrix plot is a good tool to use.

🖑 **Graphs ➤ Scatter...** This time, select Matrix and click **Define...** Now select the variables Out-of-state tuition, Instructional expenditures per student, and Avg combined SAT score as the matrix variables.

In the resulting plot (shown below), we see scatterplots relating each pairing of these three variables. In the first row, both of the graphs have tuition on the *y* axis; in the first column, tuition forms the *x* axis. You should recognize the plot of tuition vs. instructional expenditures. **What do you see in the plot of tuition versus SAT scores?**

Matrix plot for Tuition-Out

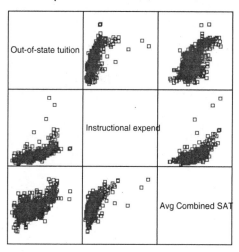

Prepared by Jane Nash & Robert Carver

The matrix plot allows us to look at several bivariate relationships at one time. In this example, though, we are hypothesizing a *multivariate* relationship: tuition depends jointly on instructional expenditures and SAT scores. Rather than think of a regression line in a two-dimensional graph, we need to think of a regression *plane* in three-dimensional space. Algebraically, we are hypothesizing a model that looks like this:

$$\text{Tuition}_i = \beta_0 + \beta_1 \text{Expenditure}_i + \beta_2 \text{SAT}_i + \varepsilon_i$$

To help visualize what that relationship might look like, we need to add a dimension to our scatterplot. SPSS lets us do so as follows:

🖱 **Graphs ➤ Interactive ➤ Scatterplot...** In the familiar dialog box (next page), first click on the button marked **2-D Coordinate**, and select **3-D Coordinate** instead. Then drag Out-of-state tuition to the vertical axis, Average combined SAT score to the diagonal axis, and Instructional expenditure per student to the horizontal axis. In the Fit tab, select Regression, and then title the graph.

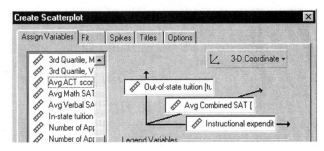

The interactive graph displays a three-dimensional graph of the data, along with an estimated regression plane and equation. Initially, you are looking along one edge of the plane. To see more of the plane, and how the points scatter around it, we continue on.

Double-click anywhere on the graph. The graph will open, along with some controls you have not seen before, shown on the facing page. Now for some fun.

In the 3-D Controls window, click on the Rotation Cursor. You'll see the cursor change to the hand shape shown here. While pressing and holding the left mouse button, slide your mouse steadily from left to right. Then release the mouse button.

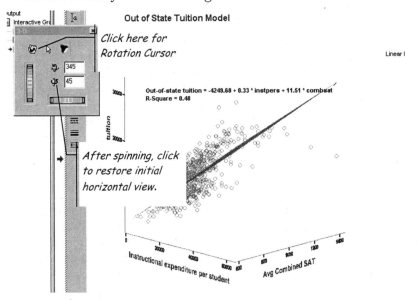

When done properly, your graph will rotate in space, revealing the "cloud" of points and their relationship to the estimated plane. Click the left mouse button once more to stop the spinning. This may take a few tries, but keep experimenting with it. It can truly help you to see what is happening with this multiple regression model.

Now we are ready to estimate our model. We do this as before, adding a new independent variable.

> 🖱 **Analyze ➤ Regression ➤ Linear...** We want to add a second variable—Average combined SAT score [combsat]—to the list of independent variables. Also, among the Statistics..., be sure to request Descriptives.

In the Output Viewer, first note that adding a variable has reduced the number of cases available for analysis from $n = 1244$ in the first regression to $n = 747$ in this one—a loss of about 40% of the data! This often happens when there is some missing data. In this instance, many schools did not report SAT scores. Therefore, although we have two of the three variables for those schools, they are dropped from the regression. We can only use cases for which we have a complete set of data. This is a consideration in selection of a multiple regression model.

Correlations

		Out-of-state tuition	Instructional expenditure per student	Avg Combined SAT
Pearson Correlation	Out-of-state tuition	1.000	.637	.610
	Instructional expenditure per student	.637	1.000	.619
	Avg Combined SAT	.610	.619	1.000
Sig. (1-tailed)	Out-of-state tuition	.	.000	.000
	Instructional expenditure per student	.000	.	.000
	Avg Combined SAT	.000	.000	.
N	Out-of-state tuition	747	747	747
	Instructional expenditure per student	747	747	747
	Avg Combined SAT	747	747	747

Below the Descriptive Statistics you will find the correlation matrix (shown above), which reports the correlation between each pairing of the three variables. The upper portion of the table, labeled Pearson Correlation, reports all of the correlations; the correlation coefficient for a pair of variables appears at the intersection of the corresponding row and column. For example, tuition and expenditures have a correlation of

0.637, while tuition and SAT have a correlation of only 0.610. Both correlations are positive, suggesting (for example) that tuition and SAT scores rise and fall together, though imperfectly. The significance levels of the correlations and the sample sizes are also reported in the table.

It is important to recognize that these correlations refer only to the pair of variables in question, without regard to the influences of other variables. Above, though, we theorized that tuition varies *simultaneously* with expenditures and SAT scores. That is to say, we suspect that SAT scores affect tuition in the context of a given expenditure level. Therefore, we can't merely look at the relationship of SAT scores and tuition without taking expenditures into account. Multiple regression allows us to do just that. Let's see how to interpret the rest of the output.

Following the correlation, most of the output should look quite familiar, with only minor differences. We see two independent variables in the Variables Entered/Removed table. The model summary shows a slight improvement in adjusted r^2 (approximately .48, up from .44) and a reduction in the standard error of the estimate ($2,901, down from $3,126). Note that the coefficient table is now longer than before.

Coefficients[a]

Model		Unstandardized Coefficients		Standardized Coefficients	t	Sig.
		B	Std. Error	Beta		
1	(Constant)	-4249.675	947.058		-4.487	.000
	Instructional expenditure per student	.332	.027	.420	12.500	.000
	Avg Combined SAT	11.505	1.107	.349	10.389	.000

a. Dependent Variable: Out-of-state tuition

We now have one intercept (Constant) and two slopes, one for each of the two explanatory variables. The intercept represents the value of *y* when *all* of the *x* variables are equal to zero. Each slope represents the marginal change in *y* associated with a one-unit change in the corresponding *x* variable, *if the other x variable remains unchanged.* For example, if expenditure were to increase by one dollar, and mean SAT scores were to remain constant, then tuition would increase by .332 dollars (i.e., 33 cents), on average. Look at the coefficient for Avg Combined SAT. ***What does it tell you?***

The coefficient table also reports *standardized* coefficients, or *betas* for each variable. These betas (or *beta weights*) allow us to compare the *relative importance* of each independent variable. In this case, instructional expenditures (beta = .420) have a greater impact on tuition than do SAT scores (beta = .349).

Significance Testing and Goodness of Fit

In linear regression, we tested for a significant relationship by looking at the *t* or *F*-ratios. In multiple regression, the two ratios test two different hypotheses. As before, the *t* test is used to determine if a slope equals zero. Thus, in this case, we have two tests to perform:

Expenditures	**SAT Scores**
H_0: $\beta_1 = 0$	H_0: $\beta_2 = 0$
H_A: $\beta_1 \neq 0$	H_A: $\beta_2 \neq 0$

The *t* ratio and significance level in each row of the table of coefficients tell us whether to reject each of the null hypotheses. In this instance, at the .05 level of significance, we reject in both cases, due to the very low *P*-values. That is to say, both independent variables have statistically significant relationships to tuition.

The *F*-ratio in a multiple regression is used to test the null hypothesis that all of the slopes are equal to zero:

$$H_0: \beta_1 = \beta_2 = 0 \text{ vs. } H_A: H_0 \text{ is not true.}$$

Note that the alternative hypothesis is different from saying that all of the slopes are nonzero. If one slope were zero and the other were not, we would reject the null in the F test. In the two *t* tests, we would reject the null in one, but fail to reject it in the other.

Finally, let's return to r^2, the coefficient of multiple determination. In the prior sessions, we noted that the output reports both r^2 and "adjusted" r^2. It turns out that adding any *x* variable to a regression model will tend to inflate r^2. To compensate for that inflation, we adjust r^2 to account for both the number of *x* variables in the model, and for the sample size.[1] When working with multiple regression analysis, we generally want to consult the adjusted figure. In this instance, the addition of another variable really does help to explain the variation in *y*.

In this regression, the adjusted r^2 equals .479, or 47.9%; in the simple regression model, using only expenditure as the predictor variable, the adjusted r^2 was only 44%. We would say that, by including the SAT Scores in the equation, we are accounting for an additional 3.9% (47.9 – 44) of the total variation in out-of-state tuitions.

---

[1] See your primary textbook for the formula for adjusted r^2. Note the presence of *n* and *k* (the number of independent variables) in the adjustment.

Residual Analysis

As in simple regression, we want to evaluate our models by the degree to which they conform to the assumptions concerning the random disturbance terms. That is why we requested the residual plots in the regression command.

We interpret the residual plots (see next page) exactly as we did before. In these particular graphs, you should note that the normal probability plot is very close to a 45° line, indicating that the normality assumption is satisfied. The residuals versus predicted values plot is less clear on the subject of homogeneity of variance (homoskedasticity). Rather than an even horizontal band of points, we see an egg-shaped cluster of residuals. However, in contrast to the earlier residual graph, this is an improvement. Besides explaining more variation than a simple regression model, a multiple regression model can sometimes resolve a violation of one of the regression assumptions. This is because the simple model may assign too much of the unexplained variation to ε, when it should be attributed to another variable.

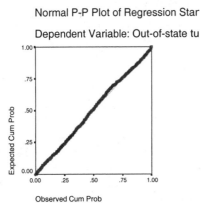

Normal P-P Plot of Regression Star

Dependent Variable: Out-of-state tu

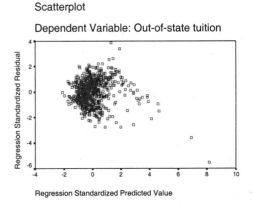

Scatterplot

Dependent Variable: Out-of-state tuition

Adding More Variables

This model improved the simple regression. Let's see if we can improve this one further by adding another variable. Suppose we hypothesize further that a school's tuition structure is also influenced by its other sources of income. Another variable in our data file is the % of alumni who contribute $ [alumcont]. Let us add this variable to our model as well.

⌐͡ **Analyze ➤ Regression ➤ Linear...** Add % of alumni who contribute $ [alumcont] to the list of predictors.

Compare this regression output to the earlier results. **What was the effect of adding this new variable to the model?** Obviously, we have an additional coefficient and *t* ratio. **Does that t ratio indicate that the % of alumni who contribute has a significant relationship to tuition, when we control for expenditures and SAT scores? Does the sign of the coefficient make sense to you? Explain.**

What else changed? Look, in particular, at the adjusted r², the ANOVA results, and the values of the previously estimated coefficients. Can you explain the differences you see?

The addition of a new variable can also have an impact on the residuals. In general, each new model will have a new residual plot. **Examine the residual graphs, and see what you think. Do the least squares assumptions appear to be satisfied?**

Another Example

In the Moving On... questions, we'll return to our analysis of tuition. Let's see another example, using the same data file, this time including a qualitative variable in the analysis. Our concern in this problem is what admissions officers call "Admissions Yield." When you were a high school senior, your college sent acceptance letters out to many students. Of that number, many chose to attend a different school. The "yield" refers to the proportion of students who actually enroll, compared to the number admitted. In this regression model, we will concentrate on the relationship between the number of seniors a college accepts, and the number who eventually enroll. Let's look at that relationship graphically.

⌐͡ Construct a simple (non-interactive) scatterplot with Number of new students enrolled [newenrol] on the *y* axis and Number of Applications accepted [appsacc] on the *x* axis.

Since there are nearly 1,300 schools in this sample, the graph is very dense. Notice that the points are tightly clustered in the lower left of the graph, but fan out as you move to the upper right. Even in this scatterplot, you can see evidence of heterogeneity of variance, or heteroskedasticity. **Why do you think that might occur? That is, what would cause the heteroskedasticity?**

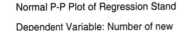 Now let's run the regression and evaluate the model. Run a regression using new enrollments as the dependent and acceptances as the sole independent variable. As before, request the residual analyses.

Look at the residual graphs. What do they suggest about the validity of the least squares assumptions in this case? These residuals violate the assumption of a normal distribution, and seem to fan out from left to right in the residuals versus predicted values plot.

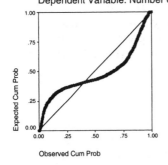

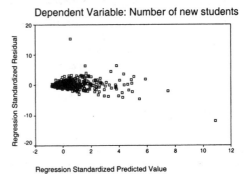

Like most regressions, this one has strengths and flaws. Examine the regression output, including the residual graphs, and evaluate the model, considering these questions:

- ***Does the coefficient of acceptances have the expected sign?***
- ***Is the relationship statistically significant?***
- ***How good is the fit?***
- ***Are there many unusual observations?***

Working with Qualitative Variables

In this regression, we find that acceptances account for a large percentage of the variation in enrollments. Another potentially important factor in explaining the differences is whether a school is publicly funded.

Whether a school is public or private is, of course, a categorical variable. All of the variables we have used in regression analysis so far are quantitative. It is possible to include a qualitative predictor variable in a regression analysis if we come up with a way to represent a categorical variable numerically.

We do this with a fairly simple trick known as a *dummy variable.* A dummy variable is an artificial binary variable which assumes arbitrarily chosen values to represent two distinct categories.[2] Ordinarily, we set up a new variable which equals 0 for one category, and 1 for the other. In this dataset, we have a variable called Public/Private School [pubpvt] which happens to equal 1 for public colleges and 2 for private colleges. We can use that variable in this case.

Let's rerun the regression, simply adding PubPvt to the independent variables. Part of the output is shown here:

Coefficients[a]

Model		Unstandardized Coefficients		Standardized Coefficients	t	Sig.
		B	Std. Error	Beta		
1	(Constant)	777.221	45.927		16.923	.000
	Number of Applications accepted	.316	.005	.804	61.476	.000
	Public/Private School	-359.687	24.162	-.195	-14.886	.000

a. Dependent Variable: Number of new students enrolled

Look at the table of coefficients. We could write the estimated regression equation like this:

Newenroll = 777.221 + .316 Appsacc − 359.687 PubPvt

Think for a moment about the meaning of the coefficient of the public/private variable. For a public college or university, PubPvt = 1, and this equation becomes:

Newenroll = 777.221 + .316 Appsacc − 359.687(1)
= 417.534 + .316 Appsacc

On the other hand, for a private school, the equation is:

Newenroll = 777.221 + .316 Appsacc − 359.687(2)
= 57.847 + .316 Appsacc

Take a moment and look very closely at the two equations. They have the exact same slope, but different intercepts (417.5 vs. 57.8). In other words, we are looking at two parallel lines whose intercepts differ

[2] It is possible to represent multinomial variables using several dummy variables. Consult your primary textbook or instructor.

by about 360 students. The impact of the dummy variable, introduced into the equation in this way, is to alter the *intercept* for the two different categories.

Now that we know what the estimated equation is, let's go on to evaluate this particular model, as we did earlier.

- **Do the estimated coefficients have the expected signs?**
- **Are the relationships statistically significant?**
- **How good is the fit?**
- **Are there many unusual observations?**
- **Are the residuals normally distributed?**
- **Are the residuals homoskedastic?**

A New Concern

Suppose we wanted to try to improve this model even further, and hypothesize that, other things being equal, tuition charges might influence the admissions yield. This seems eminently reasonable: High school seniors choosing between two equally competitive private schools might select the less expensive one. Indeed, if we run a regression including in-state tuition, we see the following coefficients table showing a significant negative coefficient for the new variable. This means that higher tuitions lead to lower yields, other things being equal.

Coefficients[a]

Model		Unstandardized Coefficients		Standardized Coefficients		
		B	Std. Error	Beta	t	Sig.
1	(Constant)	636.207	52.776		12.055	.000
	Number of Applications accepted	.332	.005	.823	62.372	.000
	Public/Private School	-224.306	38.293	-.121	-5.858	.000
	In-state tuition	-1.36E-02	.003	-.082	-4.308	.000

a. Dependent Variable: Number of new students enrolled

Suppose we wanted to embellish the model by adding both in-state and out-of-state tuition to the equation. Let's do so, and rerun the regression with out-of-state tuition added to the list of predictors. By now, you should be able to make this change.

Look at this regression output, and take special note of the total variation explained, the standard error, and the other regression coefficients and test statistics. *Neither tuition* slope appears to be statistically significant at the .05 level! What is happening here?

This is an illustration of a special concern in multiple regression: *multicollinearity.* When two or more of the predictor variables are highly correlated in the sample, the regression procedure cannot determine which predictor variable concerned is associated with changes in y. In a real sense, regression cannot "disentangle" the individual effects of each x. In this instance, the culprits are in-state and out-of-state tuition, which have a correlation coefficient of .928.

This is the root of the problem, and with this sample, can only be resolved by eliminating one of the two variables from the model. Which should we eliminate? We should be guided both by theoretical and numerical concerns: we have a very strong theoretical reason to believe that tuition belongs in the model, and in-state tuition has the stronger numerical association with enrollments. Given a strong theoretical case, it is probably wiser to retain in-state tuition and omit the out-of-state.

Moving On...

Now apply the techniques learned in this session to the following questions. Each question calls upon you to devote considerable thought and care to the analysis. As you write up your results, be sure that you offer a theoretical explanation for a relationship between the dependent variable and each independent variable you include in a model. Also address these specific questions:

- Are signs of coefficients consistent with your theory?
- Are the relationships statistically significant?
- Do the residuals suggest that the assumptions are satisfied?
- Is there any evidence of a problem with multicollinearity?
- How well does the model fit the data?

Colleges

1. In the session, we have found three variables that help to estimate new enrollment. Let's see if we can expand the model to do a more complete job. Your task is to choose one more variable from among the following, and add it to the regression model:
   - Top10
   - FacTerm

- AlumCont
- GradRate

You may choose any one you wish, provided you can explain how it logically might affect new enrollments, once acceptances, tuition, and public/private are accounted for. Then run the regression model including the new variable, and evaluate the regression in comparison to the one we have just completed in the session.

2. In the session we also developed a multiple regression model for out-of-state tuition. Develop your own model for in-state tuition, using as many of the variables in the file as you wish, provided you can explain why each one should be in the model.

States

3. Develop a multiple regression model to estimate the number of fatal injury accidents using as many other available variables in the data file as you see fit (excluding Auto Accident Fatalities). One variable you must include is the Blood Alcohol Content Threshold (BAC), which represents the legal definition of driving while intoxicated in each state. Do states which permit a higher threshold have more traffic fatalities, other things being equal?

Bodyfat

4. Develop a multiple regression model to estimate the body fat percentage of an adult male (FatPerc), based on one or more of the following easily-measured quantities:
   - Age (years)
   - Weight (pounds)
   - Abdomen circumference (in cm)
   - Chest circumference (in cm)
   - Thigh circumference (in cm)
   - Wrist circumference (in cm)

You should refer to a matrix plot and/or correlation matrix to help select variables. Your model may contain any or all of the variables listed here. Also, discuss possible logical problems with using a linear model to estimate body fat percentage.

Sleep

5. Develop a multiple regression model to estimate the total amount of sleep (Sleep) required by a mammal species, based on one or more of the following variables:
    *   Body weight
    *   Brain weight
    *   Lifespan
    *   Gestation

    You should refer to a matrix plot and/or correlation matrix to help select variables. Your model may contain any or all of the variables listed here. Also, discuss possible logical problems with using a linear model to estimate sleep requirements.

Labor

6. Develop a multiple regression model to estimate the mean number of new weekly unemployment insurance claims. Select variables from this list, based on theory, and the impact of each variable in the model.
    *   Civilian labor force (A0M441)
    *   Ratio of index of newspaper ads to number unemployed (A0M060)
    *   Number of people unemployed (A0M037)
    *   Civilian unemployment rate (A0M043)

    You should refer to a matrix plot and/or correlation matrix to help select variables. Your model may contain one, two, three, or all four predictor variables. As always, be sure you can explain why each independent variable is in the model.

US

7. Develop a multiple regression model to estimate aggregate personal savings. As always, be sure you can explain why each independent variable is in the model.

Utility

8. During the study period, the owners added a room to this house, and thereby increased its heating needs. Use the variable NewRoom in a multiple regression analysis (including measures of temperature) to estimate the additional number of therms per day consumed as a consequence of enlarging the house.

Impeach

9. Develop a multiple regression model to estimate the number of guilty votes cast by a senator, using as many available variables as you see fit. As always, be sure you can explain why each independent variable is in the model.

F500

10. Develop a multiple regression model to estimate profit, using as many available variables as you see fit. As always, be sure you can explain why each independent variable is in the model. NOTE: You should not incorporate any variables that implicitly include profit (e.g. earnings per share).

Nonlinear Models

Objectives

In this session, you will learn to do the following:
- Improve a regression model by transforming the original data
- Interpret coefficients and estimates using transformed data

When Relationships Are Not Linear

In our regression models thus far, we have assumed *linearity*; that is, that y changes by a fixed amount whenever an x changes by one unit, other things being equal. The linear model is a good approximation in a great many cases. However, we also know that some relationships probably are not linear. Consider the "law of diminishing returns" as illustrated by weight loss. At first, as you reduce your calories, pounds may fall off quickly. As your weight goes down though, the rate at which the pounds fall off may diminish.

In such a case, x and y (calories and weight loss) are indeed related, but *not in a linear fashion*. This session provides some techniques that we can use to fit a *curve* to a set of points. Our basic strategy with each technique will be the same. We will attempt to find a function whose characteristic shape approximates the curve of the points. Then, we'll apply that function to one or more of the variables in our worksheet, until we have two variables with a generally linear relationship. Finally, we'll perform a linear regression using the transformed data. We will begin by using an artificial example. In SPSS, open the data file called **Xsquare**.

A Simple Example

Let's begin with a very familiar nonlinear relationship between two variables, in which y varies with the square of x. The formal model (known as a quadratic model) might look like this:

$$y = 3x^2 + 7$$

In fact, the **Xsquare** file is an artificial set of data that reflects that exact relationship. If we plot y versus x, and then plot y versus xsquare, the graphs look like this:

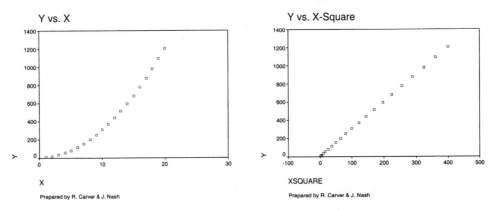

In the left graph, we see the distinct parabolic shape of a quadratic function. y increases each time that x increases, but does so by an *increasing* amount. Clearly, x and y are related, but just as clearly, the relationship is *curvilinear*. In the second graph, we have a perfectly straight line that is an excellent candidate for simple linear regression. **If you were to run a regression of y on xsquare, what would the slope and intercept be?** (Do it, and check yourself.)

This illustrates the strategy we noted above: When we have a curved relationship, we'll try to find a way to transform one or more of our variables until we have a graph which looks linear. Then we can apply our familiar and powerful tool of linear regression to the *transformed* variables. *As long as we can transform one or more variables and retain the basic functional form of y as a sum of coefficients times variables*, we can use linear regression to fit a curve. That is the basic idea underlying the next few examples. SPSS provides several ways to approach such examples; we'll explore two of them.

Some Common Transformations

In our first artificial example, we squared a single explanatory variable. As you may recall from your algebra and calculus courses, there are many common curvilinear functions, such as cubics, logarithms, and exponentials. In this session, we'll use a few of the many possible transformations just to get an idea of how one might create a mathematical model of a real-world relationship.[1]

Let's begin with an example from Session 15 (and about 400 years ago): Galileo's experiments with rolling balls. Recall that the first set of data plotted out a distinct curve:

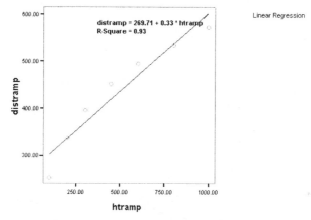

The greater the starting height, the further the ball rolled, but the increased horizontal roll diminishes as heights increase. A straight line is not a bad fit ($r^2 = .93$), but we can do better with a different functional form. In fact, Galileo puzzled over this problem for quite some time, until he eventually reasoned that horizontal distance might vary with the square root of height.[2] If you visualize the graph of $y = +\sqrt{x}$, it looks a good deal like the scatterplot above: It rises quickly from left to right, gradually becoming flatter.

---

[1] Selection of an appropriate model should be theory-driven, but sometimes can involve trial and error. The discussion of theoretical considerations is beyond the scope of this book; consult your primary text and instructor for more information.

[2] For an interesting account of his work on these experiments, see David A. Dickey and Arnold, J. Tim, "Teaching Statistics with Data of Historic Significance," *Journal of Statistics Education* v. 3, no. 1 (1995).

🖱 Open the file called **Galileo**.

🖱 **Transform ➤ Compute...** As shown here, create a new variable (SqrtHt) equal to the *square root* of HtRamp.

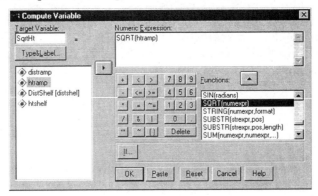

🖱 **Graphs ➤ Interactive ➤ Scatterplot...** DistRamp is on the *y* axis, and SqrtHt on the *x* axis. Include the regression line, generating this graph:

Galileo's Ramp & Ball Experiment

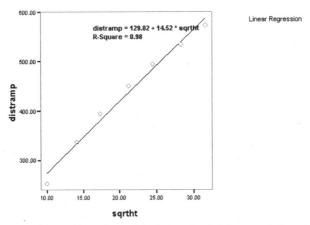

This transformation does tend to straighten out the curved line. What is more, it makes theoretical sense that friction will tend to "dampen" the horizontal motion of a ball rolled from differing heights. Other functional transformations may align the points more completely, but don't make as much sense, as we'll see later.

Now that we've straightened the dots, let's take a look at the resulting estimated regression equation that is included with the graph:

$$DistRamp = 129.02 + 14.52\sqrt{RampHt}$$

The intercept means that a ball rolled from a height of 0 *punti* would roll about 129 *punti*, and that the distance would increase by 14.5 *punti* each time the square root of height increased by one *punto*. The fit is excellent ($r^2 = .98$), and if we use the regression procedure to compute significance tests (not shown here) we find that they strongly suggest a statistically significant relationship between the two variables.

Using the equation, we can estimate the distance of travel by simply substituting a height into the model. For instance, if the initial ramp height were 900 *punti*, we would have:

$$DistRamp = 129.02 + 14.52\sqrt{900}$$
$$= 129.02 + 14.52(30)$$
$$= 564.62 \ punti$$

Note that we must take care to transform our *x* value here to compute the estimated value of *y*. Our result is calculated using the square root of 900, or 30.

Like any regression analysis, we must also check the validity of our assumptions about ε. The next example includes that analysis, as well as another curvilinear function.

Another Quadratic Model

Nonlinear relationships crop up in many fields of study. Let's consider the relationship between aggregate personal savings and aggregate personal income, which you may have seen in an earlier **Moving On...** question. We might expect that savings increase as income increases, but not necessarily in a linear fashion. Open the **US** file.

This example will also introduce a new SPSS command for handling nonlinear estimation. Let's start with the simple linear model.

 Graphs ➤ Interactive ➤ Scatterplot... Create a scatterplot with aggregate savings on the vertical axis and aggregate income on the horizontal. Include a regression line.

As you can see in the resulting graph (next page), the points arc around the fitted line, and r^2 is only .84. Let's run this regression using a new command that allows us to specify a nonlinear model as well.

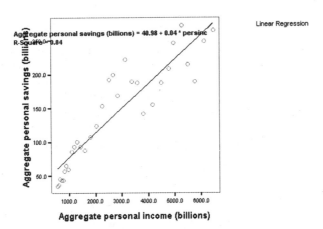

Savings vs. Income

🖰 **Analyze ➤ Regression ➤ Curve Estimation** As shown in this dialog box, select savings and income as the variables, and specify both a Linear and a Quadratic model. Then click **Save....**

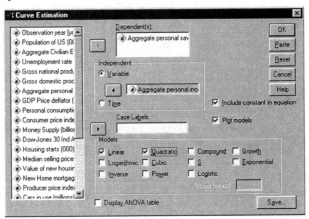

🖰 In the Save Variables area of the Save dialog box, select Predicted Values and Residuals. Click **Continue**, and then **OK**.

This command has several effects. In the Output Viewer, you see the results of two regressions as well as a graph. In the Data Editor, SPSS has created four new variables: predicted values and residuals for the two regressions. Let's start by examining the residuals.

In previous regressions, the Regression procedure automatically gave us the standard residual graphs. In this instance, we'll need to generate our own graphs. Switch to the Data Editor, and scroll to the right. Notice the variables called fit_1 and err_1. These are the "fitted" (predicted) values and "errors" (residuals) from the linear model. We can plot them to check our assumptions.

   🖱 **Graphs ➤ P-P...** This command generates a normal probability plot. Select the variable Error for PERSSAV...MOD_1 LINEAR [err1], click Standardize values under Transform, and click **OK**.

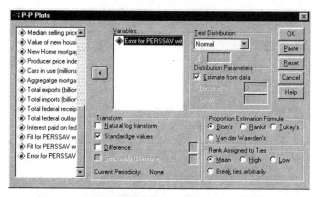

Examine the normal probability plot. ***Does it suggest normally distributed disturbances?*** Now let's plot the residuals versus predicted values.

   🖱 **Graphs ➤ Scatter...** Define a simple scatterplot with the first predicted values variable (fit_1) on the *x* axis and the first residuals variable (err_1) on the *y* axis.

   What do you see in this residual plot? Does the linearity assumption seem reasonable?
   Look back at the original scatterplot of income and savings. Can you visualize the left half of an inverted parabola? Such a pattern might fit this *quadratic* model:

$$Savings_i = \beta_0 + \beta_1 Income_i + \beta_2 Income_i^2 + \varepsilon_i$$

Now return to the portion of your output labeled Curve Fit. When we ran the Curve Estimation procedure, we asked for both a linear and a quadratic model. This command is a shortcut way to run a regression with transformed data. The Curve Estimation dialog box offers eleven different models. To interpret the output from this command, we need to know the functional forms of these models. Here is a summary of several of the simple alternatives used in this session:[3]

Model	Form of Equation
Logarithm	$y = \beta_0 + \beta_1 \ln(x)$
Quadratic	$y = \beta_0 + \beta_1 x + \beta_2 x^2$
Cubic	$y = \beta_0 + \beta_1 x + \beta_2 x^2 + \beta_3 x^3$

Curve Fit

```
MODEL:   MOD_1.
_
Independent:   PERSINC

  Dependent Mth   Rsq   d.f.       F  Sigf      b0      b1      b2

   PERSSAV  LIN   .838    30  154.82   .000  40.9798   .0364
   PERSSAV  QUA   .887    29  113.46   .000   1.7822   .0739  -6.E-06
```

```
The following new variables are being created:
  Name          Label
  FIT_1         Fit for PERSSAV with PERSINC from CURVEFIT, MOD_1 LINEAR
  ERR_1         Error for PERSSAV with PERSINC from CURVEFIT, MOD_1 LINEAR
  FIT_2         Fit for PERSSAV with PERSINC from CURVEFIT, MOD_1 QUADRATIC
  ERR_2         Error for PERSSAV with PERSINC from CURVEFIT, MOD_1 QUADRATI
```

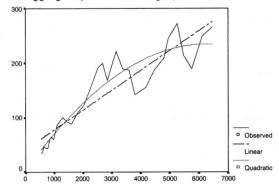

Aggregate personal savings (billions)

Aggregate personal income (billions)

---

[3] If you point your cursor at the name of a model and click the right mouse button, you'll see the equation. Also see Session 19 for additional models related to forecasting.

The output provides us with two estimated models. Using b0, b1, and b2 from the output, we can rewrite the second estimated equation like this:

$$Savings_i = 1.7822 + .0739 Income_i - .000006 Income_i^2 + \varepsilon_i$$

How does this regression equation compare to the linear model? What do the two slopes tell you? Are the results statistically significant? Has the coefficient of multiple determination (adjusted r^2) improved? How do these residuals compare to those in the linear model? What strengths and weaknesses do you find in these residuals?

One theoretical problem with this quadratic model is that it begins to slope down as you move to the right side of the graph; that is, it suggests that savings will decline when income rises, say, to $7,000 billion. Let's try the cubic model, which continues to increase throughout the values of income.

🖰   Return to the Curve Estimation dialog box; deselect Linear and Quadratic, and select Cubic.

How does the cubic model compare to the quadratic? Examine the residuals; are they better in this model or in the quadratic model?

As you can see, the cubic model is not perfect, but from this short example, you can get a sense of how the technique of data transformation can become a very useful tool. Let's turn to another example, using yet another transformation.

A Log-Linear Model

For our last illustration, we'll return to the household utility dataset (**Utility**). As in prior labs, we'll focus on the relationship between gas consumption (gaspday) and mean monthly temperature (meantemp).

You may recall that there was a strong negative linear relationship between these two variables when we performed a simple linear regression (r^2 was .864). There were some problems with the regression, though. The plot of Residuals vs. Predicted values suggested some nonlinearity.

When you think about it, it makes sense that the relationship can't be linear over all possible temperature values. In the sample, as the temperature rises, gas consumption falls. In the linear model, there must be some temperature at which consumption would be negative, which obviously cannot occur. A better model would be one in which consumption falls with rising temperatures, but then levels off at some point, forming a pattern similar to a natural logarithm function. The natural log of temperature serves as a helpful transformation in this case; that is, we will perform a regression with gaspday as *y*, and *ln*(meantemp) as *x*. Such a model is sometimes called a log-linear model. The **Curve Estimation** command does the job nicely. First, in the Data Editor, open **Utility**; don't save the changes made in **US**.

🖱 **Analyze ➤ Regression ➤ Curve Estimation...** This time, the dependent variable is gaspday and the independent is meantemp. Select both the linear and logarithmic models, and save the predicted values and residuals.

The estimated logarithmic equation turns out to be:

$$\text{gaspday} = 40.5974 - 9.3380 \ ln(\text{meantemp})$$

What are the strengths and weaknesses of the logarithmic regression? What is your interpretation of the residual analysis? In the linear model, a one-degree increase in temperature is associated with a decrease of about 0.22 therms. How do we interpret the slope in the logarithmic model?

Apply the same logic as we always have. The slope is the marginal change in *y*, given a one-unit change in *x*. Since *x* is the natural log of temperature, the slope means that consumption will decrease 9.338 therms when the *log* of temperature increases by 1. The key here is that one-unit differences in the natural log function are associated with ever-increasing changes in temperature as we move up the temperature scale.

In Session 15, we used the linear regression model to predict gas consumption for a month in which temperature was 40 degrees. Suppose we want to do that again with the transformed data. We can't simply substitute the value of 40, since *x* is no longer temperature, but rather it is the natural logarithm of temperature. As such, we must substitute *ln*(40) into the estimated equation. Doing so will yield an estimate of gas consumption. **What is the estimated consumption for a month with a mean temperature of 40 degrees?**

In the simple linear model, we obtained a negative consumption estimate for a mean temperature of 75°. **Estimate consumption with this new model, using the ln(75). Is this result negative also?**

Adding More Variables

We are not restricted to using simple regression or to using a single transformed variable. All of the techniques and caveats of multiple regression still apply. In other words, one can build a multiple regression model that involves some transformed data and other nontransformed variables as well. The **Curve Estimation** command limits us to a single independent variable, but we have seen how to use the **Compute** command to transform any variable we like, permitting us to use multiple independent variables.

In addition, we can transform the dependent variable. This requires additional care in the interpretation of estimates, because the fitted values must be *untransformed* before we can work with them.

Moving On...

Galileo

1. Return to the data from the first Galileo experiment (first two columns), and using the **Curve Estimation** command, fit quadratic and cubic models. Discuss the comparison of the results.

2. Use the two new models to estimate the horizontal roll when a ball is dropped from 1,500 *punti*. Compare the two estimates to an estimate using the square root model. Comment on the differences among the three estimates, select the estimate you think is best, and explain why you chose it.

3. Fit a curvilinear model to the data in the third and fourth columns. Use both logic and statistical criteria to select the best model you can formulate.

Bodyfat

4. Compare and contrast the results of linear, quadratic, and logarithmic models to estimate body fat percentage using abdomen circumference as an independent variable. Evaluate the logic of each model as well as the residuals and goodness-of-fit measures.

Sleep

5. Compare and contrast the results of linear, quadratic, and logarithmic models to estimate total sleep hours using gestation period as an independent variable. Evaluate the

logic of each model as well as the residuals and goodness-of-fit measures.

Labor

6. Compare and contrast the results of linear, quadratic, and logarithmic models to estimate the ratio of help-wanted ads to the number of unemployed persons, using civilian unemployment rate as an independent variable. Evaluate the logic of each model as well as the residuals and goodness-of-fit measures.

7. Compare and contrast the results of linear, quadratic, and logarithmic models to estimate the average weekly new unemployment claims using the ratio of help-wanted ads to the number of unemployed persons as an independent variable. Evaluate the logic of each model as well as the residuals and goodness-of-fit measures.

Colleges

8. Build a model of out-of-state tuition using the natural logarithm of expenditure per student. Compare the results of the log-linear model with those of the simple linear model. Be sure to analyze residuals as well as the other regression outputs.

9. [Advanced] We can build a multiple regression model using transformed data. In your Data Editor, compute two new variables representing the natural logs of expenditures per student and average combined SAT scores. Then, using the **Linear Regression** command, estimate a model using Out-of-State tuition as the dependent variable and *ln*(expenditure) and *ln*(combsat) as the independents. Compare the results of this multiple regression model with those of the model in the previous question.

Utility

10. One variable in this file is called Heating Degree Days [hdd]. It equals the sum of daily mean temperature deviations below a base temperature of 65° F. Thus, a month with a high value for HDD was very cold. Using gaspday as the dependent

variable and hdd as the independent, estimate and compare the linear, quadratic, cubic, and logarithmic models. Which seems to be the best?

Bowling

These are results from a bowling league. Each persons bowls a "series" consisting of three "strings" (maximum score = 300 per string).

11. Suppose we want to know if we can predict a bowler's series total based on the score of his or her first string. Compare a linear model to another model of your choice, referring to all relevant statistics and graphs.

Output

12. Construct linear, cubic, and logarithmic models with the Index of industrial production as the dependent and durables production as the independent variable. Compare the strengths and weaknesses of the models.

13. Construct linear, cubic, and logarithmic models with durables production as the dependent and nondurables production as the independent variable. Compare the strengths and weaknesses of the models.

14. Construct linear, cubic, and logarithmic models with the consumer goods production as the dependent and durables production as the independent variable. Compare the strengths and weaknesses of the models.

Session 19

Basic Forecasting Techniques

Objectives

In this session, you will learn to do the following:
- Identify common patterns of variation in a time series
- Make and evaluate a forecast using Moving Averages
- Make and evaluate a forecast using Trend Analysis

Detecting Patterns over Time

In the most recent sessions, our concern has been with building models which attempt to account for variation in a dependent, or response, variable. These models, in turn, can be used to estimate or predict future or unobserved values of the dependent variable.

In many instances variables behave predictably over time. In such cases, we can use *time series forecasting* to predict what will happen next. There are many time series techniques available; in this session, we will work with two of them. The SPSS Base system includes a few time series tools. The SPSS Trends module offers extensive and powerful methods, but that is beyond the scope of this book.

Recall that a *time series* is a sample of repeated measurements of a single variable, observed at regular intervals over a period of time. The length of the intervals could be hourly, daily, monthly; what is most important is that it be *regular*. We typically expect to find one or more of these common idealized patterns in a time series, often in combination with one another:

- *Trend:* General upward or downward pattern over a long period of time, typically years. A time series showing no trend is sometimes called a *stationary time series.* For example, common stock prices have shown an upward trend for many years.

- *Cyclical variation:* Regular pattern of up-and-down waves, such that peaks and valleys occur at regular intervals. Cycles emerge over a long period of years. The so-called "Business Cycle" may be familiar to some students.
- *Seasonal variation:* Pattern of ups and downs within a year, repeated in subsequent years. Most industries have some seasonal variation in sales, for example.
- *Random, or irregular, variation:* Movements in the data, which cannot be classified as one of the preceding types of variation.

Let's begin with some real-world examples of these patterns. Be aware that it is rare to find a real time series which is a "pure" case of just one component or another. We'll start by opening the **US** file.

Some Illustrative Examples

All of the variables in this file are measured annually. Therefore, we cannot find seasonal variation here.

🖰 **Graphs ➤ Sequence...** In the dialog box (see below), we'll select the variables Population of US (000) [pop], Money Supply (billions) [m1], Housing starts (000) [starts][1], Unemployment rate (%) [unemprt], and New Home mortgage rate [nhmort]. Select One chart per variable; this will create five graphs.

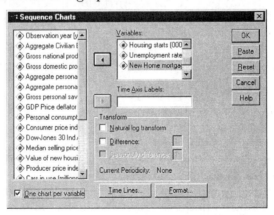

---

[1] Housing starts refers to the number of new homes on which construction began in the year.

The first graph shows the population of the United States, and one would be hard-pressed to find a better illustration of a linear trend. During the period of the time series, population has grown by a nearly constant number of people each year. It is easy to see how we might extrapolate this trend.

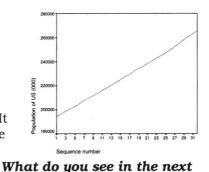

What do you see in the next graph (m1)? There is also a general trend here, but it is *nonlinear*. If you completed the session about nonlinear models, you might have some ideas about a functional form which could describe this curve. In fact, as we'll see later, this graph is a typical example of growth which occurs at a constant *percentage* rate, or exponential growth.

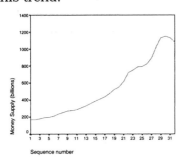

The third graph (Housing starts) is a rough illustration of *cyclical* variation, combined with a moderate negative trend. Although the number of starts increases and decreases, the general pattern is downward, with peaks and valleys spaced fairly evenly. Notice also that a slight downward trend is evident in the series. This is what we mean when we say that the components sometimes appear in combination with one another.

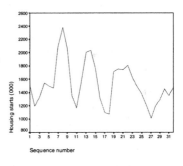

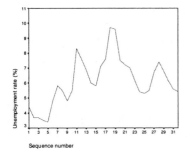

The fourth graph, showing the unemployment rate during the period, has unevenly spaced peaks, and the upward trend visible on the left side of the graph appears to be flattening or even declining on the right side. The irregularities here suggest a sizable erratic component. This graph also illustrates another way various patterns might combine in a graph.

Finally, the mortgage rates graph shows almost entirely irregular movement. The pattern is not easily classified as one of the principal components noted earlier.

To see seasonal variation, we return to the **Utility** file, with home heating data from New England. Open that file now in the Data Editor.

🖑 **Graph ➤ Sequence...** Select the variable MeanTemp, and then choose Observation date for the Time Axis Labels.

We have added some vertical lines at each January to help visualize the *seasonal variation* in the data. **Comment on the extent to which this graph shows evidence of seasonal variation. What accounts for the pattern you see?**

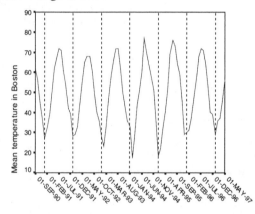

Observation date

We can exploit patterns such as these to make predictions about what will happen in future observations. Any attempt to predict the future will of necessity yield imperfect forecasts. Since we can't eliminate errors in our predictions, the trick in forecasting is to *minimize* error.

Forecasting Using Moving Averages[2]

The first technique we'll explore is useful for "smoothing out" the erratic jumps of irregular movements. It is known as Moving Averages.

---

[2] In the remainder of the session, we will assume that your primary text covers the theory and formulas behind these techniques. Consult your text for questions about these issues. The SPSS Trends module offers full-featured moving average procedures.

We'll illustrate the technique using a single time series from the **Utility** file: the mean kilowatt-hours consumed per day.

Moving Averages is an appealing technique largely due to its simplicity. We generate a forecast by merely finding the mean of a few recent observations. The key analytical issue is to determine an appropriate number of recent values to average. In general, we make that determination by trial and error, seeking an averaging period that would have minimized forecast errors in the past.

Thus, in a Moving Average analysis, we select a *span* or interval, compute retrospective "forecasts" for an existing set of data, and then compare the forecast figures to the actual figures. We summarize the forecast errors using some standard statistics explained below. We then repeat the process several more times with different spans to determine which interval length would have performed most accurately in the past.

Before making any forecasts, let's look at the time series.

🖰 Construct a sequence plot of kwhpday. *Comment on any patterns or unusual features of this plot.*

🖰 Now switch back to the Data Editor.

🖰 **Transform ➤ Create Time Series...** In the dialog box, specify a Prior moving average for a span of 4 months. Then, select the variable Mean KwH consumed per day [kwhpday].

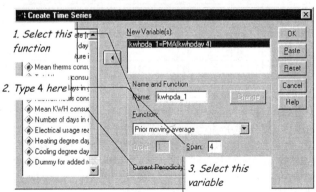

This command will create a new variable in the dataset. For each month (starting in month 5), the value of the new variable will equal the simple average of the prior 4 months. Switch back to the Data Editor to see what the command has done.

Now let's graph the original time series and the 4-month prior Moving Average using a sequence plot.

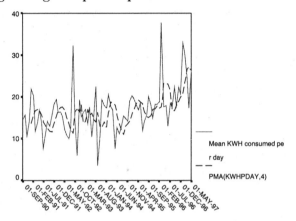

Observation date

The Moving Average line follows the pattern of the observed data, but does not reach the extreme peaks and valleys of the original data. By averaging four values at a time, we "smooth out" the original curve.

If we want to forecast a future period, say June 1997[3], we have only to compute the mean of the preceding four months (see Data Editor shown here). In this instance, that works out to 24.625.[4]

	month	days	meantemp	gaspday	therms	gasdays	kwh	kwhpda
77	01-JAN-97	31	29	9.30	307	33	1115	32.80
78	01-FEB-97	28	36	9.70	283	29	853	30.50
79	01-MAR-97	31	37	7.90	230	29	713	24.60
80	01-APR-97	30	46	5.80	171	29	498	17.20
81	01-MAY-97	31	56	3.20	104	32	838	26.20

This is how we forecast a 4-month Moving Average. But how do we know if this is a reliable forecast? Perhaps a 5-month (or 6-, or 12-month) Moving Average would give us a better result. In general, we are looking for a Moving Average length which minimizes forecast error. Let's generate a measure of forecast error, and then run another analysis, and select the one with the better error statistics.

---

[3] It may seem odd to speak of "forecasting" a figure from the past. Any projection beyond the current set of data is considered a forecast.

[4] (30.5 + 24.6 + 17.2 + 26.2)/4 = 24.625.

We will measure the forecast errors in this way: For each forecast, we'll compute the difference (or "error") between the actual electricity usage and the forecast usage. We'll square those differences, and find their mean. In other words, we'll compute the Mean Squared Error (MSE) for this particular Moving Average series. Switch to the Data Editor.

🖱 **Transform ➤ Compute...** Create a new variable called ErrSqr, equal to (kwhpday-kwhpda_1)**2.

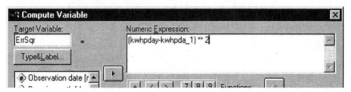

🖱 **Analyze ➤ Descriptive Statistics ➤ Descriptives...** Find the mean of ErrSqr.

The MSE for a 4-month Moving Average is 32.24. Let's repeat the process for a 6-month Moving Average and find the MSE for that series.

🖱 Repeat the process for a 6-month span. This requires that you Create a Time Series (kwhpda_2, span = 6), compute errsqr2 using the new time series, and find the mean of errsqr2.

🖱 Create a sequence plot showing the original data and both Moving Average models. *Comment on the comparison of the three lines.*

Which of the two averaging periods is better in terms of MSE? What is the forecast for June 1997 using the better model?

Forecasting Using Trend Analysis

It is clear from the graphs that there has been a slight upward trend in the Carver family's electricity usage. Let's try to model that trend, using the Curve Estimation function that we saw in Session 18.[5] First, we'll use a simple linear model, and generate two forecasts beyond our observed dataset of 81 months.

🖱 **Analyze ➤ Regression ➤ Curve Estimation...** Select Mean KwH per day as the dependent variable, and Time as the independent.

---

[5] If you did not do Session 18, you should read through it now.

Specify a Linear model, and check Display ANOVA table. Then click **Save....** Complete the Save dialog box as shown here:

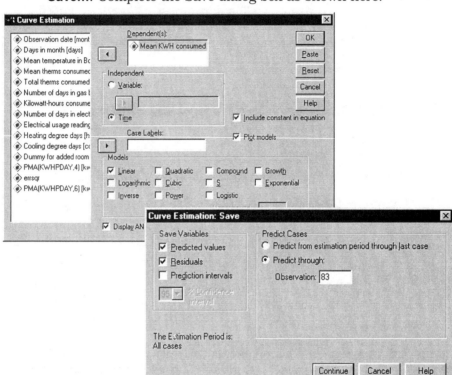

This command generates a regression analysis, computes predicted and residual values, and plots both the original data and the trend line. The regression results (see facing page) indicate good significance tests, but an unimpressive fit (adjusted r^2 is about .17) and the standard error of the estimate is 5. The standard deviation of y, not shown here, is 5.6, indicating that the model is not much of an improvement over simply using the mean value of y for forecasting. The residual plots (not shown; generated separately) are acceptable. A glance at the graph shows that the linear model is naïve at best.

Notice also that the results indicate the creation of new variables plus two new cases. These new cases are the forecasts beyond the dataset. Switch to the Data Editor and scroll to the bottom of the file. ***What are the predicted values of electricity usage for June and July of 1997?***

```
Dependent variable.. KWHPDAY            Method.. LINEAR

Listwise Deletion of Missing Data

Multiple R              .42329
R Square                .17918
Adjusted R Square       .16879
Standard Error         5.07138

            Analysis of Variance:
            DF     Sum of Squares      Mean Square
Regression   1          443.5203       443.52029
Residuals   79         2031.7940        25.71891

F =      17.24491      Signif F =  .0001

------------------- Variables in the Equation -------------------
Variable            B          SE B        Beta        T    Sig T

Time            .100081      .024100     .423293     4.153  .0001
(Constant)    12.617654     1.137490                11.093  .0000

The following new variables are being created:

    Name         Label

    FIT_1        Fit for KWHPDAY from CURVEFIT, MOD_15 LINEAR
    ERR_1        Error for KWHPDAY from CURVEFIT, MOD_15 LINEAR

2 new cases have been added.
```

Since the linear model appears to be inadequate, let's try some nonlinear trends. In addition to the linear model, the **Curve Estimation** command includes several other functional forms that are useful in forecasting. Some commonly used models are the following:

Model	Form of Equation	Alternate Form
Logarithm	$y = \beta_0 + \beta_1 \ln(x)$	
Quadratic	$y = \beta_0 + \beta_1 x + \beta_2 x^2$	
Cubic	$y = \beta_0 + \beta_1 x + \beta_2 x^2 + \beta_3 x^3$	
Compound	$y = \beta_0(\beta_1^x)$	$\ln(y) = \ln(\beta_0) + x\ln(\beta_1)$
S	$y = e^{(\beta_0 + (\beta_1/x))}$	$\ln(y) = \beta_0 + \beta_1/x$
Growth	$y = e^{(\beta_0 + (\beta_1 \cdot x))}$	$\ln(y) = \beta_0 + \beta_1 \cdot x$
Exponential	$y = \beta_0 \cdot e^{(\beta_1 \cdot x)}$	$\ln(y) = \ln(\beta_0) + \beta_1 \cdot x$

🖰 Return to the Curve Estimation dialog box and deselect Linear. We will try three common models for a time series that tends to increase: S, Growth, and Exponential.

None of these analyses is particularly satistfying, but which of the resulting analyses (including Linear) is best? Why did you select that one?

Another Example

Before leaving the topic of Trend Analysis, let's apply the technique to an example from the **US** data file. Reopen that file now. We'll take a look at the annual Interest paid on federal debt (billions) [fedint]. The U.S. government each year pays interest on the national debt.

🖰 **Analyze ➤ Regression ➤ Curve Estimation...** The dependent variable is Interest paid on federal debt (billions) [fedint]. The independent variable is Time. As before, display the ANOVA table, save predicted and residual values, and make two predictions (through period 34). Select the Linear, Quadratic, and Growth models.

Which of the three models seems to do the best job of forecasting Interest on the federal debt? Explain your choice.

Moving On...

Using the techniques presented in this lab, answer the following questions.

US

1. Perform a Trend Analysis on the variable M1, using Linear, Quadratic, Exponential, and S models. Which model seems to fit the data best? Why might that be so?

2. Find another variable in this dataset which is well-modeled by the same function as M1. Why does this variable have a shape similar to M1?

3. Create a sequence plot of the variable representing new home mortgage rates. Which components of a time series do you see in the graph? Explain.

4. Perform a 3- and a 5-year Moving Average analysis of new home mortgage rates. Use the better of the two models to forecast mortgage rates for 1997.

Output

These questions focus on A0M047, which is an *index* of total industrial production in the United States, much like the Consumer Price Index is a measure of inflation.

5. Which of the four components of a time series can you see in this series? Explain.

6. Generate 9- and 12-month Moving Averages for this series. Which averaging length would you recommend for forecasting purposes, and why? Generate one forecast using that averaging span.

7. Select a Trend model which you feel is appropriate for this variable, and generate one forecast.

8. Compare your various forecasts. If you wished to forecast this variable, which of the predictions would you rely upon, and why?

Utility

9. Generate a three-month Moving Average forecast for the variable kwhpday. Given the pattern in the entire graph, explain why it may be unwise to rely on this forecast.

10. Why might it be unwise to use a few recent months of data to predict next month's usage? What might be a better approach?

Eximport

This file contains monthly data about selected U.S. exports and imports.

11. Look at a sequence plot of Domestic agricultural exports [A0M604]. Comment on what you see.

12. Compare a 6- and 12-month Moving Average analysis for this variable. Which averaging span is better, and why?

13. Compare linear, quadratic, and exponential trend models for this variable. Which model is superior, and why?

EuropeC

Data in this file are real annual consumption as a percentage of GDP for 15 European countries.

14. Construct a time series graph for Denmark, and comment on noteworthy features of the graph. What approach would you take to forecasting a 1991 figure for Denmark, and why?

15. Construct a time series graph for Greece, and comment on noteworthy features of the graph. What approach would you take to forecasting a 1991 figure for Greece, and why?

16. Construct a time series graph for Italy, and comment on noteworthy features of the graph. What approach would you take to forecasting a 1991 figure for Italy, and why?

Labor

Data in this file are various monthly labor market measures for the United States from 1948 through 1996.

17. Perform an appropriate Trend Analysis to make a single forecast for labor market participation among women. Explain your choice of models, and comment on the results.

18. Perform an appropriate Trend Analysis to make a single forecast for labor market participation among men. Explain your choice of models, and comment on the results.

19. Perform an appropriate Trend Analysis to make a single forecast for labor market participation among teenagers (16–19). Explain your choice of models, and comment on the results.

20. Compare and contrast your findings regarding the labor market participation among women, men, and teenagers during this period of U.S. history.

Chi-Square Tests

Objectives

In this session you will learn to do the following:
- Perform and interpret a chi-square goodness-of-fit test
- Perform and interpret a chi-square test of independence

Qualitative vs. Quantitative Data

All of the tests we have studied thus far have been appropriate exclusively for quantitative variables. The tests presented in this session are suited for analyzing qualitative (e.g., nominal variables) or discrete quantitative variables, and the relationship between two such variables. The tests fall into two categories: goodness-of-fit tests and tests of independence.

Chi-Square Goodness-of-Fit Test

The chi-square goodness-of-fit test uses frequency data from a sample to test hypotheses about population proportions. That is, in these tests we are assessing how well the sample data fits the population proportions specified by the null hypothesis.

Let's put ourselves in the position that Gregor Mendel, the famous geneticist, found himself in during the 1860s. He was conducting a series of experiments on peas and was interested in the heredity of one particular characteristic—the texture of the pea seed. He observed that pea seeds are always either smooth or wrinkled.

Mendel determined that smoothness is a dominant trait. In each generation, an individual pea plant receives genetic material from two parent plants. If either parent plant transmitted a smoothness gene, the resulting pea seed would be smooth. Only if both parent plants transmitted wrinkled genes would the offspring pea be wrinkled (called a recessive trait).

If the parent peas each have one smooth and one wrinkled gene (SW), then an offspring can have one of four possible combinations: SS, SW, WS, or WW. Since smoothness dominates, only the WW pea seed will have a wrinkled appearance. Thus, the probability of wrinkled offspring in this scenario is just .25.

Over a number of experiments, Mendel assembled data on 7,324 second-generation hybrid pea plants, whose parent plants were both SW. Mendel found that 5,474 of these offspring plants were smooth and the rest (1,850) were wrinkled.

We can use a chi-square goodness-of-fit test to determine whether Mendel's observed data fits what would be expected by the inheritance model described above (i.e., a 25% chance of wrinkled offspring and a 75% chance of smooth offspring when the parent plants are both SW). This model will serve as the null hypothesis for our test. We can state this formally as follows:

H_0: $p_{wrinkled}$ = .25, p_{smooth} = .75
H_A: H_0 is false

🖱 We start by entering Mendel's data into a new Data Editor file. Click on the Variable View tab to define two variables. Name the first variable texture. For simplicity's sake, let's just use most of the default variable attributes, and create value labels. Specify that the Measure type (last column to the right) is Nominal.

🖱 Under the Values column, click on the three-dotted gray box, and complete the Value Labels dialog box as shown here.

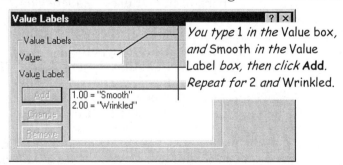

🖱 Name the second variable frequenc. Then return to the Data View and type in the data as shown here.

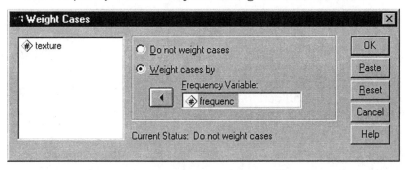

Before running our chi-square goodness-of-fit test, we need to let SPSS know that the numbers in the second column are frequency counts, not scores, from a sample. We do so as follows:

🖱 **Data ➤ Weight Cases...** Select Weight cases by and frequenc as the Frequency Variable so your dialog box looks like this:

Now let's find out whether Mendel's sample data follow the inheritance model described above.

🖱 **Analyze ➤ Nonparametric Tests ➤ Chi-Square...** (See next page.) Select texture as the Test Variable. Under Expected Values, first type in .75 and then click on **Add**. Next, type in .25 and then click on **Add**. Your dialog box should now look like the one shown on the next page.

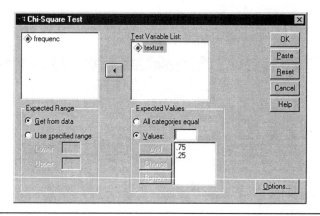

> 🖳 The order that you type in the values predicted by the null
> hypothesis (the **Expected Values** in this dialog box) is very
> important. The first value of the list must correspond to the
> lowest value of the test variable. In our analysis, the first value of
> the list (.75) corresponds to the proportion predicted for smooth,
> which is coded "1," the lowest value for the texture grouping
> variable. Similarly, the last value of the list must correspond to the
> highest group value of the test variable (e.g., .25 for wrinkled,
> which is coded "2" for the texture grouping variable).

Your output will consist of several parts. Let's begin with the
frequency data at the top of your output.

TEXTURE

	Observed N	Expected N	Residual
smooth	5474	5493.0	-19.0
wrinkled	1850	1831.0	19.0
Total	7324		

The column labeled Observed N is simply the frequency counts
that Mendel observed in his sample. Notice the column labeled Expected
N. These are the expected frequencies that would be predicted by the
null hypothesis. For example, if the null hypothesis were true in
Mendel's data, we would expect 25% of the 7,324 total offspring plants
to be wrinkled (.25 x 7,324 = 1,831). Confirm for yourself the expected
frequency for the smooth peas.

The chi-square statistic is calculated by comparing these observed frequencies (f_o) and expected frequencies (f_e). Specifically, the formula is as follows:

$$\chi^2 = \sum_{i=1}^{k} \frac{(f_{oi} - f_{ei})^2}{fei}$$

where

f_{oi} is the observed frequency of the ith category, and
f_{ei} is the expected frequency of the ith category, and
k is the number of categories.

When there is a large discrepancy between the observed and expected frequencies, we reject the null hypothesis. Alternatively, when there is a small discrepancy between the observed and expected frequencies, we fail to reject the null hypothesis. ***Do these frequencies seem discrepant to you?***

Test Statistics

	TEXTURE
Chi-Square[a]	.263
df	1
Asymp. Sig.	.608

a. 0 cells (.0%) have expected frequencies less than 5. The minimum expected cell frequency is 1831.0.

We see in this output a very small chi-square value (.263) with a very small significance associated with it (.608). What this means in terms of observed and expected frequencies is that they are very close to each other and so we fail to reject the null hypothesis. Thus, we conclude what Mendel observed in his data very closely matches (fits) the inheritance model he was testing.

Chi-Square Test of Independence

The chi-square statistic is also useful for testing whether there is a statistical relationship between two qualitative or discrete variables. The logic of this test is much the same as in the goodness-of-fit test. We start with a null hypothesis that predicts a set of frequencies. We then compare the frequencies observed in the data to the frequencies predicted by the null hypothesis. If there is a large discrepancy between the observed and expected frequencies, we reject the null hypothesis.

For instance, in the **Student** dataset (open it now), we have a binary variable which represents whether the student owns a car (owncar), and another represents whether the student is a resident living on campus or a commuter (res). It is reasonable to think these two variables might be related, with commuters having a greater need for an automobile. We can use the chi-square test of independence to determine whether the data support or refute this prediction between the variables.

The null hypothesis in the chi-square test of independence is always that the two variables are not related (i.e., they are independent of one another). For our data, the null hypothesis would be that ownership and residency are not related to each other.

Let's suppose that one-tenth of the students are commuters. If the null hypothesis is correct (commuting and car ownership are unrelated), then we should find that one-tenth of the car owners commute and an equal one-tenth of the noncar owners also commute. We understand that the sample results might not show precisely one-tenth for each group, and therefore anticipate some small departures from the values specified by the null hypothesis. If, however, the discrepancies are sufficiently large between the observed frequencies of the sample data and expected frequencies predicted by the null hypothesis, we will reject the null hypothesis.

Let's find out whether car ownership and residency are related to each other.

🖰 **Analyze ➤ Descriptive Statistics ➤ Crosstabs...** Select owncar as the Row and res as the Column, as shown below.

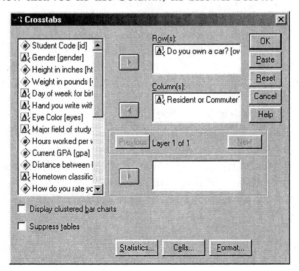

🖱 Click on **Statistics...** and choose Chi-Square.

🖱 Click on **Cells...** and under Counts select Expected (keep Observed, the default, selected as well). This allows us to see both the observed and the expected frequencies.

The output from this analysis consists of several parts. Let's first look at the cross-tabulation.

Do you own a car? * Resident or Commuter? Crosstabulation

			Resident or Commuter?		
			C	R	Total
Do you own a car?	N	Count	6	72	78
		Expected Count	17.8	60.2	78.0
	Y	Count	44	97	141
		Expected Count	32.2	108.8	141.0
Total		Count	50	169	219
		Expected Count	50.0	169.0	219.0

The rows of the table represent car ownership and the columns represent residency (Commuters in the first column, Residents in the second). Each cell of the table contains an observed frequency (Count) and an expected frequency (Expected Count). For example, look in the top left cell. There were only 6 commuters who do not own cars. Under the null hypothesis, in a total sample of 219 students, with 50 commuters and 78 noncar owners, 17.81 were expected in that cell.

Let's find out whether there is a large enough discrepancy between the observed and expected frequencies in the whole table to reject the null hypothesis.

Chi-Square Tests

	Value	df	Asymp. Sig. (2-sided)	Exact Sig. (2-sided)	Exact Sig. (1-sided)
Pearson Chi-Square	15.759[b]	1	.000		
Continuity Correction[a]	14.453	1	.000		
Likelihood Ratio	17.952	1	.000		
Fisher's Exact Test				.000	.000
N of Valid Cases	219				

a. Computed only for a 2x2 table

b. 0 cells (.0%) have expected count less than 5. The minimum expected count is 17.81.

Although this output reports several test statistics that can be used to evaluate these data, we will focus on the Pearson chi-square statistic, which is equal to 15.759 with a significance equal to .000. Thus, we reject the null hypothesis and conclude that car ownership and residency are not independent of each other. In other words, these two variables *are* significantly related.

Another Example

The chi-square test of independence comes with one *caveat*—it can be unreliable if the expected count in any cell is less than five. In such cases, a larger sample is advisable. For instance, let's look at another possible relationship in this dataset. One question on the survey asks students to classify themselves as "below average," "average," or "above average" drivers. Let's ask if that variable is related to gender.

 Analyze ➤ Descriptive Statistics ➤ Crosstabs... Select How do you rate your driving? [drive] as the Row and gender as the Column. The rest of the dialog box remains as is.

Take note of the warning that appears at the bottom of the Chi-Square Tests table in the output (shown on next page). In this instance, the test results may be unreliable due to the uncertainty created by the two cells with very low expected counts. Strictly speaking, we should not come to any inference based upon this sample data. As a description of the sample, though, it is appropriate to note that the men in this sample seem to have higher opinions of their own driving than the women do of theirs.

How do you rate your driving? * Gender Crosstabulation

			Gender		Total
			F	M	
How do you rate your driving?	Below Average	Count	4	4	8
		Expected Count	3.5	4.5	8.0
	Average	Count	61	45	106
		Expected Count	46.2	59.8	106.0
	Above Aveage	Count	30	74	104
		Expected Count	45.3	58.7	104.0
Total		Count	95	123	218
		Expected Count	95.0	123.0	218.0

Chi-Square Tests

	Value	df	Asymp. Sig. (2-sided)
Pearson Chi-Square	17.727[a]	2	.000
Likelihood Ratio	18.033	2	.000
N of Valid Cases	218		

a. 2 cells (33.3%) have expected count less than 5. The minimum expected count is 3.49.

This is an important warning to heed.

Moving On...

Use the techniques of this session to respond to these questions. *For each question, explain how you come to your statistical conclusion, and suggest a real-world reason for the result.*

Census90

1. Is the ability to speak English independent of gender?

2. Is job-seeking ("looking for work") independent of gender?

Student

3. Is seat-belt usage independent of car ownership?

4. Is seat-belt usage independent of gender?

5. Is travel outside of United States independent of gender?

6. Is seat-belt usage independent of familiarity with someone who has been struck by lightning?

Mendel

This file contains summarized results for another one of Mendel's experiments. In this case, he was interested in four possible combinations of texture (smooth/wrinkled) and color (yellow/green). His theory would have predicted proportions of 9:3:3:1 (i.e., smooth yellow most common, one-third of that number smooth green and wrinkled yellow, and only one-ninth wrinkled green).[1] The first column of the

---

[1] Heinz Kohler, *Statistics for Business and Economics* 3rd ed. (New York: HarperCollins, 1994), p. 458–459.

dataset contains the four categories, and the second column contains the observed frequencies.

7. Perform a goodness-of-fit test to determine whether these data refute Mendel's null hypothesis of a 9:3:3:1 ratio. The ratio translates into theoretical proportions of .56, .19, .19, and .06 as the Expected Values in the chi-square dialog box.

8. Renowned statistician Ronald A. Fisher reanalyzed all of Mendel's data years later, and concluded that Mendel's gardening assistant may have altered the results to bring them into line with the theory, since *each one* of Mendel's many experiments yielded chi-square tests similar to this and the one shown earlier in the session. Why would so many consistent results raise Fisher's suspicions?

Salem

This file contains the data about the residents of Salem Village during the witchcraft trials of 1692.

9. Are the variables proparri and accuser independent?

10. Are the variables proparri and defend independent?

GSS94

11. According to our 1990 Massachusetts Census file, 60.3% of respondents 18 and older were married, 13.7% were widowed, 6.1% were divorced, 1.5% were separated, and 18.3% were never married. Do these proportions fit the Marital status sample data in this file? (NOTE: For this test, do *not* weight cases; simply run the test, entering the hypothetical proportions as listed here.) Comment on what you find.

12. Is there a statistically significant difference in the way men and women answer the following questions?
    •   Should marijuana be legalized?
    •   Should a woman be able to have an abortion if there is a chance of a serious birth defect?
    •   Ever had sex with someone other than spouse while married?
    •   Who shops for the groceries?

Helping

This file contains data collected by two students who became interested in helping behavior during their Research Methods course. More specifically, they wanted to explore how the gender of the potential helper and the gender of the victim impact the likelihood of receiving help. The study was conducted in the stairwell of the college library, where either a male or a female was seen by a passer-by picking up books that he or she had dropped. The dependent variable in this study was whether the passer-by asked if help was needed or started to help pick up the books. Prior research has shown that males and females prefer to help in different ways; males tend to help in problem-solving tasks whereas females tend to help in a nurturing way. This particular situation was chosen since it did not seem to fit clearly in either type of classic male or female helping scenario.

13. Does the gender of the potential helper (gendsubj) have a significant impact on whether a person (male or female) received help? In other words, were males or females more inclined to help when they approached the individual whose books had dropped? Are the results what you would have predicted?

14. Does the gender of the victim (gendvict) make a significant difference in whether he or she received help?

Nonparametric Tests

Objectives

In this session, you will learn to do the following:

- Perform and interpret a Mann-Whitney U test for comparing two independent samples
- Perform and interpret a Wilcoxon Signed Ranks test for comparing two related samples
- Perform and interpret a Kruskal-Wallis H test for comparing two or more independent samples
- Perform and interpret a Spearman's Rank Order correlation

Nonparametric Methods

Many of the previous sessions have illustrated statistical tests involving population parameters such as μ and which often require assumptions about the population distributions (e.g., normality and homogeneity of variance). Sometimes we cannot assume normality, and sometimes the data we have do not lend themselves to computing a mean (for example, when the data are merely ordinal). The techniques featured in this session are known as *nonparametric methods*, and are applicable in just such circumstances. In particular, we will primarily use them when we cannot assume that our sample is drawn from a normal population.

As a general rule, it is preferable to use a parametric test over a nonparametric test, since parametric tests tend to be more discriminating and powerful. However, nonparametric tests should be used when the data do not meet the basic assumptions that underlie the statistical procedure (e.g. normality or homogeneity of variance).

There are a wide array of nonparametric procedures available, and SPSS includes most of them, but this session will focus on a few of the more common elementary techniques.

Mann-Whitney U Test

The Mann-Whitney U test is the nonparametric version of the independent samples *t* test. More specifically, we use this test when we have two independent samples and can assume they are drawn from populations with the same shape, although not necessarily normal. The Mann-Whitney U test can be used with ordinal, interval, or ratio data.

The basic procedure behind this test involves combining all the scores from both groups and ranking them. If a real difference between the two groups exists, then the values or scores from one sample will be clustered at one end of the entire rank order and the scores from the other sample should be clustered at the other end. Alternatively, if there is no difference between the groups, then the scores from both samples will be intermixed within the entire rank order. The null hypothesis is that the scores from the two groups are not systematically clustered and thus there is no difference between the groups.

Recall the data from the Salem Witchcraft Trials. In that dataset, we have information about the taxes paid by each individual in Salem Village. We can distinguish between those who were supporters of the Village minister, Reverend Parris, and those who were not. In this test, we will hypothesize that supporters and nonsupporters paid comparable taxes.

Before performing the test, let's check the normality of these distributions. Open the data file called **Salem**.

🖱 **Graphs ➤ Interactive ➤ Histogram...** Select Tax paid as the variable for the horizontal axis and proparri as the Panel Variable. Click on the Histogram tab and select Normal curve. Title and put your name on your histogram in the Titles tab.

Our histograms are on the facing page. Here we see two skewed distributions of comparable shape, which are clearly not normal.

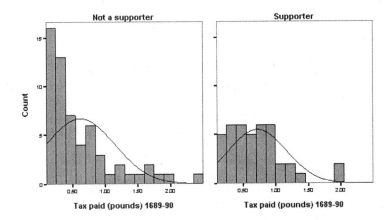

These data are more appropriately analyzed using the nonparametric Mann-Whitney U test than the parametric independent samples *t* test. Let's proceed with the Mann-Whitney U test now by doing the following:

Analyze ➤ Nonparametric Tests ➤ 2 Independent Samples... The Test Variable List is Tax paid and the Grouping Variable is proparri. There are several types of tests to choose from in this dialog box but we will stay with the default test, the Mann-Whitney U.

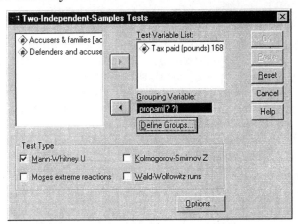

Click on **Define Groups...** and the following dialog box appears prompting us to designate which group is 1 and which is 2.

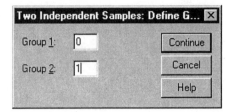

We will call the nonsupporters Group 1 (they are coded 0 in the data file) and we'll call the supporters Group 2 (they are coded 1).

Mann-Whitney Test

Ranks

	ProParris	N	Mean Rank	Sum of Ranks
Tax paid (pounds) 1689-90	Not a supporter	59	45.67	2694.50
	Supporter	41	57.45	2355.50
	Total	100		

Test Statistics[a]

	Tax paid (pounds) 1689-90
Mann-Whitney U	924.500
Wilcoxon W	2694.500
Z	-2.004
Asymp. Sig. (2-tailed)	.045

a. Grouping Variable: ProParris

All 100 scores were assigned a rank (1, 2, 3 ... 100) by Tax paid. The mean ranking of Parris' supporters was 57.45.

The first part of this output shows some summary information about the two groups; notice the sum of the ranks for each group. The Mann-Whitney U statistic is equal to 924.5 with a significance (*P*-value) equal to .045. Thus, we conclude that there is a significant difference in the amount of taxes paid by pro- and anti-Parris residents. **Which group paid more in taxes? How do you know?**

Wilcoxon Signed Ranks Test

The Wilcoxon Signed Ranks test is the nonparametric version of the paired samples *t* test. In particular, we use this test when we have repeated measures from one sample, but the parent population is not necessarily normal in shape. Like the Mann-Whitney U test, the Wilcoxon Signed Ranks test can be used with ordinal, interval, or ratio data.

The data for this test consist of the difference scores from the repeated measures. These differences are then ranked from smallest to largest by their absolute value (i.e., without regard to their sign).

If a real difference between the two measures or treatments exists, then the difference scores will be consistently positive or consistently negative. On the other hand, if there is no difference between the treatments, then the difference scores will be intermixed evenly. The null hypothesis is that the difference scores are not systematic and thus there is no difference between the treatments.

Ordinarily, a nonparametric test is performed when an assumption of a parametric test has been violated, usually the normality assumption. For purposes of an example, we will perform a Wilcoxon Signed Ranks test on some heart rate data from the **BP** data file. Specifically, we will compare heart rate during a cold water immersion task and heart rate during a mental arithmetic task. Please note, these data do not violate normality but are being used for purposes of illustration.

🖰 Open the data file **BP**.

🖰 **Analyze ➤ Nonparametric Tests ➤ 2 Related Samples...** Select heart rate cold pressor [hrcp] and heart rate mental arithmetic [hrma] by clicking on them in the list of variables. They will then appear in the Current Selections box. Next, click on the arrow key and the following should appear in the Test Pair(s) List:

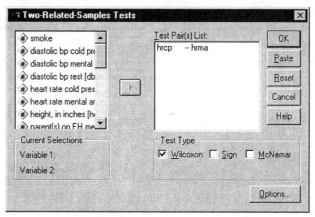

There are several tests to choose from in this dialog box but we will stick with the default test, the Wilcoxon Signed Ranks.

Wilcoxon Signed Ranks Test

Ranks

		N	Mean Rank	Sum of Ranks
heart rate mental arithmetic - heart rate cold pressor	Negative Ranks	107[a]	93.58	10013.50
	Positive Ranks	68[b]	79.21	5386.50
	Ties	3[c]		
	Total	178		

a. heart rate mental arithmetic < heart rate cold pressor

b. heart rate mental arithmetic > heart rate cold pressor

c. heart rate cold pressor = heart rate mental arithmetic

Test Statistics[b]

	heart rate mental arithmetic - heart rate cold pressor
Z	-3.447[a]
Asymp. Sig. (2-tailed)	.001

a. Based on positive ranks.

b. Wilcoxon Signed Ranks Test

The first part of the output summarizes the results. Negative ranks refer to subjects whose heart rate during mental arithmetic was less than their heart rate during the cold water immersion task. Positive ranks are those subjects whose heart rate during mental arithmetic was the higher of the two tasks. In this case there were three subjects whose heart rates were the same in these tasks. Notice the sum of the ranks for the Negative Ranks compared to the Positive Ranks. The Wilcoxon Signed Ranks statistic, converted to a z-score, is equal to –3.447 with a significance (P-value) equal to .001. Thus, we conclude that heart rate *does* change significantly during the cold water immersion task compared to the mental arithmetic task. ***During which task is heart rate higher? How do you know?***

Kruskal-Wallis H Test

The Kruskal-Wallis H test is the nonparametric version of the one-factor independent measures ANOVA. We use this test if we have more than two independent samples and can assume they are from populations with the same shape, although not necessarily normal. The Kruskal-Wallis H test can be used with ordinal, interval, or ratio data.

Like the Mann-Whitney U test, the Kruskal-Wallis H test ranks all of the observed scores. If differences among the groups exist, then scores from the various samples will be systematically clustered in the entire rank order. Alternatively, if there are no differences between the groups, the scores will be intermixed within the entire rank order. The null hypothesis states that there are no differences among the groups, and therefore the scores will not cluster systematically.

Let's look at the data from Dr. Stanley Milgram's famous experiments on obedience to authority. Under a variety of conditions, subjects were instructed to administer electrical shocks to another person; in reality there were no electric shocks, but subjects believed that there were (see pages 148–149 for a detailed description of these experiments). One factor in these experiments was the proximity between the subject and the person "receiving" the shocks. Let's find out whether proximity to the victim had a significant impact on the maximum amount of shock delivered. Before performing the test, we want to check the assumption of normality. Open the **Milgram** data file.

Graphs ➤ Interactive ➤ Histogram... Select volts as the variable for the horizontal axis and exp as the Panel Variable. Superimpose a normal curve on the histograms, and title the graph.

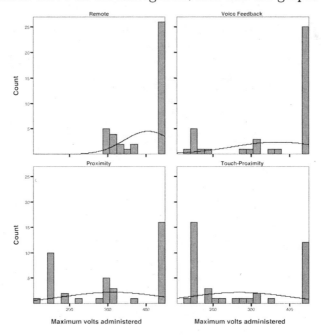

Here we see four histograms that are not normal in shape. Thus, these data are more appropriately analyzed using the nonparametric Kruskal-Wallis H test than the parametric independent measures ANOVA. Let's proceed with the Kruskal-Wallis H test now.

🖰 **Analyze ➤ Nonparametric Tests ➤ K Independent Samples...** The Test Variable List is Maximum volts administered [volts] and the Grouping Variable is exp. There are two types of tests available in this dialog box but we'll use the default, the Kruskal-Wallis H test.

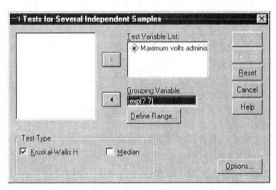

🖰 Click on **Define Range...** The following dialog box prompts us to designate the minimum and maximum values for the grouping variable. In our data file, the groups are labeled from 1 to 4.

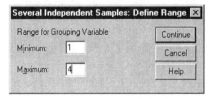

Kruskal-Wallis Test

Ranks

	Experiment	N	Mean Rank
Maximum volts administered	Remote	40	101.66
	Voice Feedback	40	91.30
	Proximity	40	71.15
	Touch-Proximity	40	57.89
	Total	160	

Test Statistics[a,b]

	Maximum volts administered
Chi-Square	24.872
df	3
Asymp. Sig.	.000

a. Kruskal Wallis Test

b. Grouping Variable: Experiment

Let's first look at the mean ranks of the various groups, which differ according to the proximity between the subject and the person he was "shocking." These rankings appear quite different but we will have to rely on the statistical test to determine whether this sense is correct. The Kruskal-Wallis statistic (chi-square) is equal to 24.872 with a significance equal to .000. Thus, we conclude that the proximity between the subject and the "victim" had a significant effect on the maximum amount of shock administered.

Spearman's Rank Order Correlation

The Spearman's Rank Order correlation is the nonparametric version of the Pearson correlation (r). Recall that the Pearson correlation measures the linear relationship between two interval or ratio variables. Sometimes, we have only ordinal data but may still suspect a linear relationship. For example, we may want to compare the income rankings of the 50 states from one year to the next.

Recall the data about the 50 states (**States**). In this dataset, we have the average wages for each state in 1993 and 1994. Suppose we wanted to determine the extent to which *rankings* of the states changed, if at all, in one year. We can explore this question by doing the following:

🖱 **Analyze ➤ Correlate ➤ Bivariate...** Select Mean wages, 1993 [pay93] and Mean wages, 1994 [pay94] as the variables. Choose Spearman's as the Correlation Coefficients and deselect Pearson, the default option, as shown here.

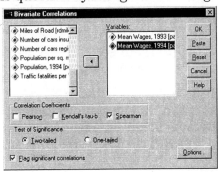

You'll find in your output that the Spearman's correlation (ρ)[1] is equal to .998, which is significantly different than zero. We can interpret this correlation as indicating a very consistent positive relationship between the rankings of the states by average wages in 1993 and 1994. As is always the case when interpreting correlations, whether they be Spearman's correlations or Pearson correlations, we have to be careful not to draw any cause and effect conclusions.

Moving On...

Apply the techniques learned in this session to the questions below. In each case, indicate which test you have used, and cite appropriate test statistics and P-values. Comment on whether a parametric test could also be used to analyze the data (e.g. check normality).

AIDS

This file contains data about the number of AIDS cases from 208 countries around the world.

1. Was there any significant change in the average *rate* of AIDS cases from 1992 to 1993? (*Hint:* Think of this as a repeated measure.)

2. Compute and interpret a Spearman's correlation coefficient for the number of cases reported in 1992 and 1994.

3. Did all six WHO regions experience roughly the same rate of cases in 1993?

Swimmer2

4. Did swimmers tend to improve between their first and last recorded time in the 50-meter freestyle?

5. Looking only at second race data, do those who compete in both events swim faster in the 50-meter freestyle than they do in the 50-meter backstroke?

---

[1] ρ is the Greek letter rho.

Census90

6. It is widely believed that gender has a significant impact on earnings. Focusing just on those persons who had wage income greater than 0, is there a significant difference between men and women? Explain.

7. Still focusing on people with positive earnings, does level of education appear to have an impact on earnings? Explain your reasoning and choice of variable.

Student

8. Does the reported number of automobile accidents vary according to students' ratings of their own driving ability?

9. Do smokers tend to drink more beer than nonsmokers?

GSS94

10. One variable in the file groups respondents into one of four age categories. Does the mean number of hours of television vary by age group?

11. Does the amount of television viewing vary by a respondent's subjectively identified social class?

Airline

This data file contains information about safety performance for airlines around the world.

12. Is there a significant difference among geographic regions in crash rates per million flight miles? Comment on what you find and offer some explanations as to what you conclude about airlines from different geographic regions.

Nielsen

This file contains Nielsen television ratings for the top 20 programs during a randomly selected week.

13. Does the rating variable vary significantly by television network?

14. Does the rank variable vary significantly by television network?

15. Closely compare the results you obtained in the previous two questions, and comment on the comparison. What can explain the differences between the results?

Tools for Quality

Objectives

In this session, you will learn to do the following:
- Create and interpret a mean control chart
- Create and interpret a range control chart
- Create and interpret a standard deviation control chart
- Create and interpret a proportion control chart
- Create and interpret a Pareto chart

Processes and Variation

We can think of any organizational or natural system as being engaged in *processes*, or series of steps that produce an outcome. In organizations, goods and services are the products of processes.

One dimension of product or service quality is the degree of process variation. That is, one aspect of a good's quality often is its *consistency*. People who are responsible for overseeing a process need tools for detecting and responding to variation in a process.

Of course, some variation may be irreducible, or at times even desirable. If, however, variation arises from the deterioration of a system, or from important changes in the operating environment of a system, then some intervention or action may be appropriate.

It is critical that managers intervene when variation represents a problem, but that they avoid unnecessary interventions which either do harm or no good. Fortunately, there are methods that can help a manager discriminate between such situations.

This lab session introduces a group of statistical tools known as *control charts*. A control chart is a time series plot of a sample statistic. Think of a control chart as a series of hypothesis tests, testing the null hypothesis that a process is "under control."

How do we define "under control?" We will distinguish between two sources or underlying causes of variation:

- *Common cause* (also called *random* or *chance*): Typically due to the interplay of factors within or impinging upon a process. Over a period of time, they tend to "cancel" each other out, but may lead to noticeable variation between successive samples. Common cause variation is always present.

- *Assignable cause* (also called *special* or *systematic*): Due to a particular influence or event, often one which arises "outside" of the process.

A process is "under control" or "in statistical control" when all of the variation is of the common cause variety. Managers generally should intervene in a process with assignable cause variation. Control charts are useful in helping us to detect assignable cause variation.

Charting a Process Mean

In many processes, we are dealing with a measurable quantitative outcome. Our first gauge of process stability will be the sample mean, $\bar{x}$. Consider what happens when we draw a sample from a process that is under control, subject only to common cause variation. For each sample observation, we can imagine that our measurement is equal to the true (but unknown) process mean, μ, plus or minus a small amount due to common causes. In a sample of n observations, we'll find a sample mean, $\bar{x}$. The next sample will have a slightly different mean, but assuming that the process is under control, the sample means should fluctuate near μ. In fact, we should find that virtually all samples fluctuate within three standard errors of the true population mean.

An $\bar{x}$ chart is an ongoing record of sample means, showing the historical (or presumed) process mean value, as well as two lines representing *control limits*. The control limits indicate the region approximately within three standard errors of the mean. SPSS computes the control limits, described further below. An example will illustrate.

Recall the household utility data (in the **Utility** file, which you should open now). In this file we have 81 monthly readings of electricity and natural gas consumption in the Carver home, as well as monthly temperature and climate data. We'll start by creating a control chart for the monthly electricity consumption, which varies as the result of seasonal changes and family activity.

The Carver family added a room and made some changes to the house, beginning roughly five years along in the dataset. It is reasonable to suspect that the construction project and the presence of additional living space may have increased average monthly electricity usage. To investigate that suspicion, we'll consider each year's data a 12-month sample, and chart the means of those samples. To do so, we must first define a new variable in the file to represent the year.

🖱 **Data ➤ Define Dates...** This command creates specialized date variables. Complete the dialog box exactly as you see here, selecting Years, months in the **Cases Are:** box, and specifying that the first case is September 1990.

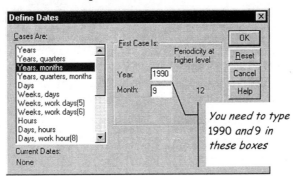

You will see a message in the Output Viewer that three new variables have been created. Switch to the Data Editor and confirm that they represent the dates of the observations in the file.

🖱 **Graphs ➤ Control...** SPSS offers several kinds of control charts appropriate to different kinds of data and different methods of organizing data. In the opening dialog box, we'll select X-Bar, R, s and the default Data Organization option (Cases are units). Click **Define**.

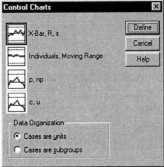

🖱 In the main dialog box (shown below), select Mean KWH consumed per day [kwhpday] as the

Process Measurement, and under Subgroups Defined by: select YEAR, not periodic. Give your graph an appropriate title.

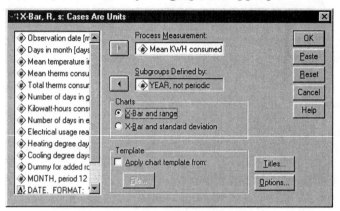

Let's take a close look at this control chart. The basic "geography" of the chart is simple. On the horizontal axis is the year; we have 8 samples of 12 observations each, collected over time. More precisely, the middle 6 years are 12 each; 1990 is based on 4 months (September through December) and 1997 on 5 months (January through May).

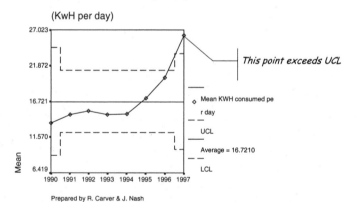

Household Electricity Usage

Prepared by R. Carver & J. Nash

The vertical axis represents the means of each sample. The solid horizontal line at 16.721, is the grand mean of all samples. The two dashed lines are the upper and lower control limits (UCL and LCL). The upper line is three standard errors above the mean, and the lower is three standard errors below[1]. For incomplete samples, such as 1990 and 1997, the control limits reflect the smaller sample size and are wider than the others.

What does this chart tell us about electricity consumption? For the first five years, the process appears to be quite stable. For the first seven, it is in control. There are no points outside of the control limits. In the sixth year, the stable pattern begins to change, rising steadily until we exceed the upper control limit in the eighth year. This suggests an assignable cause—a construction project requiring power tools, and new living spaces. **Why might usage have continued to increase? Speculate about possible explanations.**

The X-Bar chart and tests presume a constant process standard deviation throughout. Before drawing conclusions based on this chart, we should examine that assumption. What is more, sample variation is another aspect of process stability. We might have a process with a rock-solid mean, but whose standard deviation grows over time. Such a process would also require some attention. Later we'll illustrate the use of a Standard Deviation (S) chart; the S chart is appropriate for samples with five or more observations. Our samples vary in size, including one sample with $n = 4$, so we'll start with the Range (R) chart.

Charting a Process Range

The Range chart tracks the sample ranges (maximum minus minimum) for each sample. It displays a mean range for the entire dataset, and control limits computed based upon the mean sample range. We have already generated the Range chart with our prior command, so let's look at it. Our Range chart appears on the next page.

This chart is comparable in structure to the X-Bar chart. The average line represents the mean of the sample ranges. Note that it breaks at the first and last samples, reflecting the different sample sizes. The control limits are once again three standard errors from the mean. In a stable process, the sample ranges should be randomly distributed within the control limits. That seems to be the case here.

It is also important to compare the X-Bar and R charts; when a process is under control, both charts should fluctuate randomly within control limits, and should not display any obvious connections (e.g. high means corresponding to high ranges). It is for this reason that SPSS prints both charts via a single command. In this output, we see stable

---

[1] SPSS computes the standard error based on the sample standard deviation from the entire dataset.

variation but rising mean consumption. Based on the mean and Range charts, we conclude that the home owners should intercede to stabilize electricity use.

Household Electricity Usage

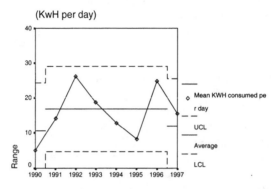

Prepared by R. Carver & J. Nash

Another Way to Organize Data

In the previous examples, the observations were in a single column. Sometimes sample data are organized so that the observations within a single sample are in a row across the data file, and each row represents the set of sample observations. To construct control charts from such data, we need only make one change in the relevant dialog boxes.

To illustrate, open the file called **EuropeY**; don't save the changes made in **Utility**. This file contains data extracted from the Penn World Tables dataset (**PAWorld**), isolating just the 14 European countries in that dataset. Each row represents one year (1960–1990), and each column a country. Each value in the worksheet is the ratio of a country's annual real per capita Gross Domestic Product to the per capita GDP of the United States, expressed as a percentage. A value of 100 means that real per capita GDP for the country was identical to that in the United States. A variety of economic and political factors presumably influence a nation's income relative to the United States; thus we may conceive of these figures as arising from an ongoing process.

      Graphs ➤ Control... We'll once again choose X-Bar, R, s, but this time, under Data Organization choose Cases are subgroups and click **Define**.

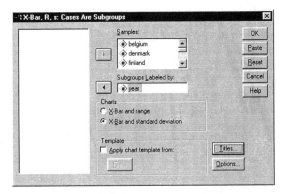

🖱 We have 14 country observations per sample. We represent this by selecting all 14 country variables as Samples. We'll label the horizontal axis by year. With larger consistent sample sizes, let's plot the standard deviation rather than the sample range by clicking on the appropriate box under Charts.

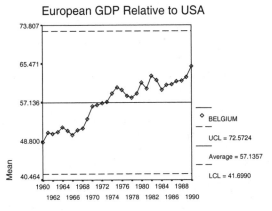

Prepared by R. Carver & J.Nash

Here the data fall well within the control limits, but we see a distinct upward drift, suggesting assignable cause variation. The mean real per capita GDP in Europe has been steadily rising relative to the United States. Due to the obvious patterns of increase, we would say that this process is *not* in statistical control, though from the perspective of the European countries, that may be just fine. **Are the standard deviations under control? Comment on the graph shown on the next page.**

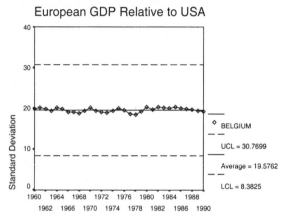

Prepared by R. Carver & J.Nash

Charting a Process Proportion

The previous examples dealt with measurable process characteristics. Sometimes, we may be interested in tracking a qualitative attribute or event in a process, and focus our attention on the frequency of that attribute. In such cases, we need a control chart for the relative frequency or sample proportion. For our example, let's consider the process of doing research on the World Wide Web.

There are a number of search engines available to facilitate Internet research. The user enters a keyword or phrase, and the search engine produces a list of Universal Resource Locators (URLs), or Web addresses relevant to the search phrase. Sometimes, a URL in the search engine database no longer points to a valid Web site. In the rapidly changing environment of the Internet, it is common for Web sites to be temporarily unavailable, move, or vanish. This can be frustrating to any Web user.

One very popular search engine is Yahoo!® which offers a feature called the Random Yahoo! Link. When you select this link, one URL is randomly selected from a massive database, and you are connected with that URL. As an experiment, we sampled twenty random links, and recorded the number of times that the link pointed to a site which either did not respond, was no longer present, or had moved. The sampling process was repeated twenty times through the course of a single day, and the results are in the data file called **Web**. Open it now.

🖱 **Graphs ➤ Control...** Choose a p, np graph in which the Cases are subunits. Complete the dialog box as shown here.

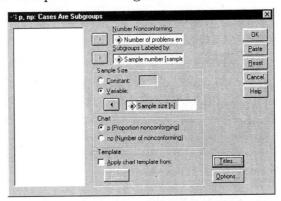

***Would you say that this process is under control? Explain
your thinking.*** Note that the lower control limit is set at 0, since we
can't have a negative proportion of problem URLs. The chart indicates
that approximately 12% of the attempted connections encountered a
problem of some kind, and that proportion remained stable through the
course of a single day.

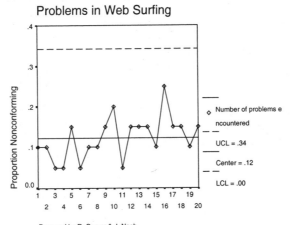

Problems in Web Surfing

Prepared by R. Carver & J. Nash

In this example, all samples were the same size (*n* = 20). Had they
been of different sizes, the control limits would have been different for
each sample.

Pareto Charts

A Pareto chart is a specialized bar chart, ordered from highest frequency to lowest, and including a cumulative relative frequency curve. It is used to focus attention on the most frequently occurring values of a nominal variable. For instance, we might have recorded the *kinds* of problem encountered during our Web-surfing experiment, and then create a Pareto chart.

We'll look at a different application, involving our data file from the Red Cross. Open the file called **Pheresis**. Pheresis refers to the process of collecting platelet cells from blood donors. To assure the quality of the blood products, the Red Cross tests every donation, measuring and counting several variables. Viable platelet donations must have a minimum concentration of platelets (which varies by collection method), minimal evidence of white blood cells, and acidity (pH level) within certain tolerances. If a donation falls outside of acceptable levels, it is more carefully analyzed and, if necessary, discarded.

Let's define a Pareto chart that plots the occurrence of problem conditions with these donations. We'll begin by analyzing a Problem Flag [flag] variable that indicates the kind of problem, if any, found in a donation.

🖰 **Graphs ➤ Pareto chart... ** As with control charts, we first indicate the kind of chart we want to make. Flag is one variable with codes representing various problems. We want SPSS to count up the frequency of each value. Thus, we specify the option for Data in Chart Are Counts or sums for groups of cases. Flag effectively groups cases by problem category.

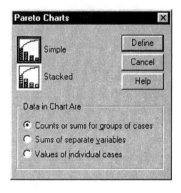

🖰 Next, as shown on the facing page, we indicate that the bars in the graph should indicate counts, and that the category (horizontal) axis variable is Problem Flag [flag].

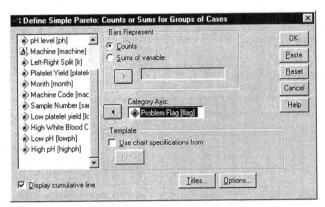

The resulting graph, shown here, indicates that the majority of donations have no problems at all. The largest number of problems involve pH levels (artificially exaggerated in this example). The data in this file are real. For the sake of this illustration, we have taken some liberties in defining the allowable tolerances for different variables. In fact, the Red Cross encounters far fewer problems than this graph would suggest.

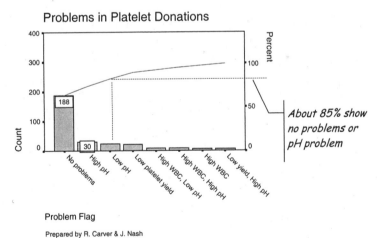

Prepared by R. Carver & J. Nash

Suppose we want to focus on the problems. One way of doing so would be to create a Pareto chart that sums up the individual problems. In our data file, we have four categorical variables representing the presence of a low platelet yield, high white blood count, low pH, or high pH. We can make the graph as follows:

🖰 **Graphs ➤ Pareto chart...** This time, we want SPSS to chart the sums of separate variables.

🖰 In the main dialog box, select the four variables just described.

How does this chart compare to the previous one? Describe what you see here.

Moving On...

Use the techniques presented in this session to examine the processes described below. Construct appropriate control charts, and indicate whether the process appears to be in control. If not, speculate about the possible assignable causes which might account for the patterns you see.

Pheresis

1. Construct mean and range control charts for platelet yield, defining subgroups by the variable Sample Number. NOTE: Sample numbers have only been assigned to three of the six pheresis machines, so you may notice that some of the data are excluded from the analysis. Discuss any noteworthy features of these control charts.

2. Use a Pareto chart to determine which of the six pheresis machines account for the majority of donations sampled in this study period. Also, which machine appears to be used least often?

Web

3. You can repeat our experiment with the Random Yahoo! Link, if you have access to a Web browser. In your browser, establish a bookmark to this URL:

   http://random.yahoo.com/bin/ryl

   Then, each time you select that bookmark, a random URL will be selected, and your browser will attempt to connect you. Tally the number of problems you encounter per 20 attempts. Each time you find a problem, also record the nature of the problem. Repeat the sampling process until you have sufficient data to construct a *p*-chart. NOTE: This process can be very time-consuming, so plan ahead.

Comment on how your *p*-chart compares to the one shown in the session.

4. Use your data about problem types to construct a Pareto chart. Which types of problems seem to be most common? Why might that be?

London1

The National Environmental Technical Center in Great Britain continuously monitors air quality at many locations. Using an automated system, the Center gathers hourly readings of various gases and particulates, twenty-four hours every day. The worksheet called **London1** contains a subset of the hourly measurements of carbon monoxide (CO) for the year 1996, recorded at a West London sensor.[2] The historical (1995) mean CO level was .669, and the historical sigma was .598.

5. Chart the sample means for these data. (*Hint:* in the first control chart dialog box, select Cases are Subgroups.) Would you say that this natural process is under control?

6. Do the sample ranges and sample standard deviations appear to be under control?

7. The filed called **London2** contains all of the hourly observations for the year (24 observations per day, as opposed to 6). Repeat the analysis with these figures, and comment on the similarities and differences between the control charts.

Labor

NOTE: To create control charts with this file, you'll need to define dates as we did with the **Utility** data. Observations in this file are also monthly, starting in January 1945.

8. The variable called A0M091 is the mean duration of unemployment as of the observation month. Using subgroups

---

[2] The file contains 325 days of data, with 6 observations at regular 4-hour intervals each day. Apparently due to equipment problems, some readings were not recorded.

defined by YEAR, develop appropriate control charts to see whether the factors affecting unemployment duration were largely of the common cause variety.

9. The variable called A0M001 represents the mean weekly hours worked by manufacturing workers in the observation month. Using subgroups defined by YEAR, develop appropriate control charts to see whether the factors affecting weekly hours were largely of the common cause variety.

10. Using the **Transform ➤ Compute** function, create a new variable called EmpRate, equal to total Civilian Employed divided by Civilian Labor Force. This will represent the *percentage* of the labor force that was actually employed during the observation month. Using subgroups defined by YEAR, develop appropriate control charts to see whether the factors affecting employment were largely of the common cause variety.

Eximport

NOTE: To create control charts with this file, you'll need to define dates as we did with the **Utility** data. Observations in this file are also monthly, starting in January 1948.

11. Using the **Transform ➤ Compute** function, create a new variable called Ratio, equal to total Exports, excluding military aid shipments divided by General Imports. Using subgroups defined by YEAR, develop appropriate control charts to see whether the factors affecting the ratio of exports to imports were largely of the common cause variety.

EuropeC

12. The data in this worksheet represent consumption as a percentage of a country's Gross Domestic Product. Each row is one year's data from fourteen European countries. Using the rows (cases) as subgroups, develop appropriate control charts to see whether the factors affecting consumption were largely of the common cause variety.

Dataset Descriptions

This appendix contains detailed descriptions of all SPSS data files provided with this book. Refer to it whenever you work with the datasets.

AIDS.SAV

Reported cases of Acquired Immune Deficiency Syndrome (AIDS) from 208 countries around the world. $n = 208$. Source: World Health Organization, 1995.
Gopher://itsa.ucsf.edu:70/11/.i/.q/.d/library/worldstats/950118 worldstats8.

Var.	Description
Country	Name of Country
WHOReg	World Health Organization Regional Office
	1 = Africa
	2 = Americas
	3 = Eastern Mediterranean
	4 = Europe
	5 = South-East Asia
	6 = Western Pacific
Case7991	Number of cases reported, 1979–1991
Case92	Number of cases reported, 1992
Rate92	Cases per 100,000 population, 1992
Case93	Number of cases reported, 1993
Rate93	Cases per 100,000 population, 1993
Case94	Number of cases reported, 1994
TotCase	Total Number of cases reported, 1979–1994

AIRLINE.SAV

Fatal accident data for airlines around the world, 1971–1999 (through June 1999). $n = 92$. Source: www.airsafe.com/airline.html.

Var.	Description
Airline	Airline name
Country	Number of countries reporting
Rate	Events per million flight miles
Events	Number of flights in which a fatality occurred
Flights	Millions of flight miles
Date	Year of last occurrence
Region	Geographic region of airline's base

ANSCOMBE.SAV

This is a dataset contrived to illustrate some hazards of using simple linear regression in the absence of a model or without consulting scatterplots. $n = 11$. Source: F. J. Anscombe, "Graphs in Statistical Analysis," *The American Statistician* v. 27, no. 1 (Feb. 1973), pp. 17–21.

Var.	Description
X1	First independent variable
Y1	First dependent variable
X2	Second independent variable
Y2	Second dependent variable
X3	Third independent variable
Y3	Third dependent variable
X4	Fourth independent variable
Y4	Fourth dependent variable

ANXIETY2.SAV

This dataset is from a study examining whether the anxiety a person experiences affects performance on a learning task. Subjects with varying anxiety and tension levels performed a learning task across four trials, and number of errors on each trial was recorded. $n = 12$. Source: SPSS Inc.

Var.	Description
Subject	Subject ID
Anxiety	Anxiety condition
Tension	Tension condition
Trial1	Number of errors in trial 1
Trial2	Number of errors in trial 2

Trial3   Number of errors in trial 3
Trial4   Number of errors in trial 4

BEV.SAV

Financial data from 91 firms in the beverage industry in fiscal 1994. $n = 91$. Sources: Company reports and data extracted from Compact Disclosure © 1996, Disclosure Inc.

Var.	Description
Company	Name of Company
Assets	Total Assets (000s of $)
Liab	Total Liabilities (000s of $)
Sales	Gross Sales
Quick	Quick Ratio ({current assets– inventory}/current liab)
Current	Current Ratio (current assets/current liab)
InvTurn	Inventory Turnover
RevEmp	Revenue per Employee
DebtEq	Debt to Equity Ratio
SIC	Principal Standard Industry Classification Code (SIC)
	2082 Malt Beverages
	2084 Wines, Brandy, & Brandy Spirits
	2085 Distilled & Blended Liquors
	2086 Bottled & Canned Soft Drinks
	2087 Flavoring, Extract & Syrup

BODYFAT.SAV

Various body measurements of 252 males, including two estimates of the percentage of body fat for each man. Sources: Penrose, K., Nelson, A., and Fisher, A., "Generalized Body Composition Prediction Equation for Men Using Simple Measurement Techniques," *Medicine and Science in Sports and Exercise* v. 17, no. 2 (1985), p. 189. Used with permission. Available via the *Journal of Statistics Education.* http://www.amstat.org/publications/jse, contributed by Prof. Roger Johnson of South Dakota School of Mines and Technology.

Var.	Description
Density	Body density, measured by underwater weighing
FatPerc	Percent body fat, estimated by Siri's equation[1]

---

[1] Siri's equation provides a standard method for estimating the percentage of fat in a person's body. Details may be found in the libstat Web site noted above, or in W. E. Siri, "Gross Composition of the Body,"

Age	Age, in years
Weight	Weight, in pounds
Height	Height, in inches
Neck	Neck circumference (cm)
Chest	Chest circumference (cm)
Abdomen	Abdomen circumference (cm)
Hip	Hip circumference (cm)
Thigh	Thigh circumference (cm)
Knee	Knee circumference (cm)
Ankle	Ankle circumference (cm)
Biceps	Biceps circumference (extended; cm)
Forearm	Forearm circumference (cm)
Wrist	Wrist circumference (cm)

BOWLING.SAV

One week's scores for the Huntsville, Alabama All Star Bowling League. $n = 27$. Source: http://fly.hiwaay.net/~jpirani/allstar/scores.html.

Var.	Description
ID	Bowler number
Game1	First game score
Game2	Second game score
Game3	Third game score
Series	Series score
Lane	Lane number

BP.SAV

Blood pressure and other physical measurements of subjects under various experimental conditions. $n = 181$. Source: Professor Christopher France, Ohio University. Used with permission.

Var.	Description
Sjcode	Subject code
Sex	Subject sex (0 = Female, 1 = Male)
Age	Subject age
Height	Height, in inches
Weight	Weight, in pounds
Race	Subject race
	1 = Amer. Indian or Alaskan Native
	2 = Asian or Pacific Islander

in *Advances in Biological and Medical Physics*, v. IV, edited by J. H. Lawrence and C. A. Tobias, (New York: Academic Press, Inc., 1956).

	3 = Black, not of Hispanic origin
	4 = Hispanic
	5 = White, not of Hispanic origin
	9 = None of the above or missing value
Meds	Taking prescription medication (0 = No, 1 = Yes)
Smoke	Does subject smoke? (0 = Nonsmoker, 1 = Smoker)
SBPCP	Systolic blood pressure with cold pressor
DBPCP	Diastolic blood pressure with cold pressor
HRCP	Heart rate with cold pressor
SBPMA	Systolic blood pressure while doing mental arithmetic
DBPMA	Diastolic blood pressure while doing mental arithmetic
HRMA	Heart rate while doing mental arithmetic
SBPREST	Systolic blood pressure at rest
DBPREST	Diastolic blood pressure at rest
PH	Parental Hypertension (0 = No, 1 = Yes)
MEDPH	Parent(s) on EH meds (0 = No, 1 = Yes)

CENSUS90.SAV

This is a random sample of raw data provided by the Data Extraction System of the U.S. Bureau of the Census. The sample contains selected responses of 982 Massachusetts residents, drawn from their completed 1990 Decennial Census forms. N= 982. Source: U.S. Dept. of Census, http://www.census.gov

Var.	Description
Age	Age of respondent (years, except as noted below)
	0 = Less than 1 year
	90 = 90 years or more
Avail	Availability for employment
	0 = n/a under 16 years old/at work/not looking
	1 = No, already has job
	2 = No, temporarily ill
	3 = No, other reasons (in school, etc.)
	4 = Yes, could have taken a job
Citizen	Citizenship (NOTE: All in sample are U.S. citizens)
	0 = Born in U.S.
	1 = Born in Puerto Rico, Guam, and outlying
	2 = Born abroad of American parents
	3 = U.S. citizen by naturalization
English	Ability to speak English
	0 = n/a less than 5 years old
	1 = Very well
	2 = Well
	3 = Not well
	4 = Not at all

Fertil   Number of children ever born
                    0 = n/a less than 15 years/male
                    1 = No child
                    2 = 1 child
                    3 = 2 children
                    .
                    .
                    .
                    12 = 11 children
                    13 = 12 or more children

Hour89   Usual hours worked per week during 1989
                    (number of hours, except as noted)
                    0 = n/a under 16 years/ did not work in 1989
                    99 = 99 or more usual hours

Hours   Hours worked last week (number of hours, except as noted)
                    0 = n/a under 16 years/did not work
                    99 = 99 or more hours last week

Income1   Wage or salary income, 1989 (dollars, except as noted)
                    0 = n/a under 16 years/none

Lang1   Language other than English at home
                    0 = n/a under 5 years old
                    1 = Yes, speaks other language
                    2 = No, speaks only English

Looking   Looking for work
                    0 = n/a or less than 16 years old/at work
                    1 = Yes
                    2 = No

Marital   Marital status
                    0 = Now married, (excluding separated)
                    1 = Widowed
                    2 = Divorced
                    3 = Separated
                    4 = Never married or under 15 years old

Trans   Means of transportation to work
                    0 = n/a, not a worker or in the labor force
                    1 = Car, truck, or van
                    2 = Bus or trolley bus
                    3 = Streetcar or trolley car
                    4 = Subway or elevated
                    5 = Railroad
                    6 = Ferryboat
                    7 = Taxicab
                    8 = Motorcycle
                    9 = Bicycle
                    10 = Walked
                    11 = Worked at home

	12 = Other method
Rearning	Total personal earnings (dollars; 0 = n/a or none)
Riders	Vehicle occupancy on way to work

        0 = n/a
        1 through 6 = number of people
        7 = 7 to 9 people
        8 = 10 or more people

Serialno	Family serial number
Sex	Sex (0 = Male, 1 = Female)
TravTime	Travel time to work (number of minutes, except as noted)

        0 = n/a; not a worker or worked at home
        1 = 99 minutes or more

YearSch	Years of school completed

        0 = n/a less than 3 years old
        1 = No school completed
        2 = Nursery school
        3 = Kindergarten
        4 = 1st, 2nd, 3rd, or 4th grade
        5 = 5th, 6th, 7th, or 8th grade
        6 = 9th grade
        7 = 10th grade
        8 = 11th grade
        9 = 12th grade, no diploma
        10 = HS graduate, diploma or G.E.D.
        11 = Some college, but no degree
        12 = Associate Degree, occupational
        13 = Associate Degree, academic
        14 = Bachelors Degree
        15 = Masters Degree
        16 = Professional Degree
        17 = Doctorate Degree

COLLEGES.SAV

Each year, *U.S. News and World Report* magazine surveys colleges and universities in the United States. The 1994 dataset formed the basis of the 1995 Data Analysis Exposition, sponsored by the American Statistical Association, for undergraduates to devise innovative ways to display data from the survey. This file contains several variables from that dataset. $n = 1302$. Source: *U.S. News and World Report,* via the *Journal of Statistics Education.* http://www.amstat.org/publications/jse. Used with permission. NOTE: Schools are listed alphabetically by state.

Var.	Description
ID	Unique identifying number

Name	Name of school
State	State in which school is located
PubPvt	Public or private school (1 = public, 2 = private)
MathSAT	Avg. math SAT score
VerbSAT	Avg. verbal SAT score
CombSAT	Avg. combined SAT score
MeanACT	Average ACT score
MSATQ1	First quartile, math SAT score
MSATQ3	Third quartile, math SAT score
VSATQ1	First quartile, verbal SAT score
VSATQ3	Third quartile, verbal SAT score
ACTQ1	First quartile, ACT score
ACTQ3	Third quartile, ACT score
AppsRec	Number of applications received
AppsAcc	Number of applications accepted
NewEnrol	Number of new students enrolled
Top10	Pct. of new students from top 10% of their HS class
Top25	Pct. of new students from top 25% of their HS class
FTUnder	Number of full-time undergraduates
PTUnder	Number of part-time undergraduates
Tuit_In	In-state tuition
Tuit_Out	Out-of-state tuition
RmBoard	Room and board costs
FacPhD	Pct. of faculty with Ph.D.'s
FacTerm	Pct. of faculty with terminal degrees
SFRatio	Student-to-faculty ratio
AlumCont	Pct. of alumni who donate
InstperS	Instructional expenditure per student
GradRate	Pct. of students who graduate within 4 years

EUROPEC.SAV

This file is extracted from the Penn World Table data. Data values are real annual consumption as a percentage of annual Gross Domestic Product. $n = 31$. Soure: See **PAWORLD.SAV** (page 297).

Var.	Description
Year	Observation year (1960–1992)
BELGIUM	Real consumption % of GDP, Belgium
DENMARK	Same, for Denmark
FINLAND	Same, for Finland
FRANCE	Same, for France
GERMANYW	Same, for W. Germany
GREECE	Same, for Greece
IRELAND	Same, for Ireland
ITALY	Same, for Italy

NETHERLA Same, for Netherlands
NORWAY   Same, for Norway
PORTUGAL Same, for Portugal
SPAIN   Same, for Spain
SWEDEN   Same, for Sweden
TURKEY   Same, for Turkey
U.K.   Same, for United Kingdom

EUROPEY.SAV

This file is extracted from the Penn World Table data. Data values are real per capita GDP relative to GDP in the United States (%; U.S. = 100%). $n = 31$. Source: See **PAWORLD.SAV** (page 297).

Var.	Description
Year	Observation year (1960–1992)
BELGIUM	Real per capita GDP as % of U.S., Belgium
DENMARK	Same, for Denmark
FINLAND	Same, for Finland
FRANCE	Same, for France
GERMANYW	Same, for W. Germany
GREECE	Same, for Greece
IRELAND	Same, for Ireland
ITALY	Same, for Italy
NETHERLA	Same, for Netherlands
NORWAY	Same, for Norway
PORTUGAL	Same, for Portugal
SPAIN	Same, for Spain
SWEDEN	Same, for Sweden
TURKEY	Same, for Turkey
U.K.	Same, for United Kingdom

EXIMPORT.SAV

Current dollar value of selected U.S. exports and imports, monthly, Jan. 1948–Mar. 1996. All numbers are millions of dollars. $n = 579$. Source: *Survey of Current Business*.

Var.	Description
Date	Month and year
A0M602	Exports, excluding military aid shipments
A0M604	Exports of domestic agricultural products
A0M606	Exports of nonelectrical machinery
A0M612	General imports
A0M614	Imports of petroleum and petroleum products

A0M616   Imports of automobiles and automobile parts

F500.SAV

Selected data about the 1996 Fortune 500 companies. $n = 500$. Source: *Time*, Inc. http://pathfinder.com/fortune. © 1996 *Time* Inc. All rights reserved. Used with permission.

Var.	Description
RRank95	Revenue ranking, 1995
RRank94	Revenue ranking, 1994
CoName	Company name
Revenue	Revenue, 1995 (millions of dollars)
RevChg	Pct. change in revenue, 1994–95
Profit	Profits (millions of dollars)
ProfChg	Pct. change in profits, 1994–95
Assets	Assets (millions of dollars)
Equity	Total Stockholders Equity (millions of dollars)
MktVal	Market value 3/15/96 (millions of dollars)
PrSale	Profits as a pct. of sales
PrAsst	Profits as a pct. of assets
PrEq	Profits as a pct. of Stockholders' Equity
EPS95	Earnings per share, 1995 (dollars)
EPSChg	Pct. change in EPS, 1994–95
Grow8595	Annual rate of growth of EPS, 1985–95 (%)
ROI95	Return to Investors, 1995 (%)
ROI8595a	Annual rate of ROI, 1985–95 (%)
Employ	Number of employees
Indust	Primary industry
State	Postal code for state of headquarters

GALILEO.SAV

Galileo's experiments with gravity included his observation of a ball rolling down an inclined plane. In one experiment, the ball was released from various points along a ramp. In a second experiment, a horizontal "shelf" was attached to the lower end of the ramp. In each experiment, he recorded the initial release height of the ball, and the total horizontal distance that the ball traveled before coming to rest. All units are *punti* (points), as recorded by Galileo. $n = 7$. Sources: Stillman Drake, *Galileo at Work* (Chicago: University of Chicago Press, 1978). Also see David A. Dickey and Arnold, J. Tim, "Teaching Statistics with Data of Historic Significance" *Journal of Statistics Education* v. 3, no. 1 (1995). Available via http://www.amstat.org/publications/jse.

Var.	Description
DistRamp	Horizontal distance traveled, ramp experiment
HtRamp	Release height, ramp experiment
DistShel	Horizontal distance traveled, shelf experiment
HtShelf	Release height, shelf experiment

GROUP.SAV

Dr. Bonnie Klentz of Stonehill College conducted an experiment to investigate subjects' perceptions of group-member contributions in a cooperative task. Controlling for group size and other factors, she recorded the actual work level for each group member and the subject's perception of coworker effort. $n = 77$. Source: Professor Klentz. Used with permission.

Var.	Description
Sub	Subject number
Gender	subject gender (1 = male, 2 = female)
Age	Subject age (in years)
Year	Year in college
	1 = Freshman
	2 = Sophomore
	3 = Junior
	4 = Senior
Grpsize	Group size (1 = Size 2, 2 = Size 3)
Subtot1	Subject's total task1
Difnext	Subject's perception of coworker

GSS8494.SAV

Selected questions from the 1984 and 1994 General Social Survey. n =4465. Source: http://www.icpsr.umich.edu/GSS/home.htm

Var.	Description
ID	Respondent's ID Number
Year	GSS year for this respondent
Polviews	Think of self as liberal or conservative?
	0 = NAP
	1 = Extremely Liberal
	2 = Liberal
	3 = Slightly Liberal
	4 = Moderate
	5 = Slightly Conservative
	6 = Conservative
	7 = Extremely Conservative

	8 = Don't Know
	9 = NA
Grass	Should marijuana be made legal?
	0 = NAP
	1 = Legal
	2 = Not Legal
	8 = Don't Know
	9 = NA

> 🖥 All of the abortion questions below are coded identically with Abany.

Abany	Should a woman be able to get an abortion for any reason?
	0 = NAP
	1 = Yes
	2 = No
	8 = Don't Know
	9 = NA
Abdefect	Abortion if strong chance of a serious defect?
Abhlth	Abortion if woman's health seriously endangered?
Abnomore	Abortion if married but wants no more children?
Abpoor	Abortion if low income- can't afford more children?
Abrape	Abortion if pregnant as a result of rape?
Absingle	Abortion if not married?
Colath	Should colleges allow anti-religionist to teach?
	0 = NAP
	4 = Allowed
	5 = Not allowed
	8 = Don't Know
	9 = NA
Colcom	Should a Communist teacher be fired?
	0 = NAP
	4 = Fired
	5 = Not fired
	8 = Don't Know
	9 = NA
Colhomo	Should colleges allow a homosexual to teach?
	0 = NAP
	4 = Allowed
	5 = Not allowed
	8 = Don't Know
	9 = NA
Colmil	Should colleges allow a militarist to teach?
	0 = NAP
	4 = Allowed
	5 = Not allowed
	8 = Don't Know

	9 = NA
Colrac	Should colleges allow a racist to teach?
	0 = NAP
	4 = Allowed
	5 = Not allowed
	8 = Don't Know
	9 = NA
Homosex	Homosexual sex relations?
	0 = NAP
	1 = Always wrong
	2 = Almost always wrong
	3 = Sometimes wrong
	4 = Not wrong at all
	5 = Other
	8 = Don't Know
	9 = NA
Fear	Afraid to walk at night in neighborhood?
	0 = NAP
	1 = Yes
	2 = No
	8 = Don't Know
	9 = NA

GSS94.SAV

Selected questions from 1994 General Social Survey. $n = 2992$.
Source: http://www.icpsr.umich.edu/GSS/home.htm.

Var.	Description
ID	Respondent's ID number
Year	GSS year for this respondent
Marital	Marital Status
	1 = Married
	2 = Widowed
	3 = Divorced
	4 = Separated
	5 = Never married
	9 = NA
Agewed	Age when first married? (age in years, except as noted)
	0 = NAP
	98 = Don't Know
	99 = NA
Age	Age of respondent (age in years, except as noted)
	98 = Don't Know
	99 = NA
Educ	Highest year of school completed (grade, except as noted)

	97 = NAP
	98 = Don't Know
	99 = NA
Degree	Highest degree of education completed by respondent
	0 = Less than High School
	1 = High School
	2 = Junior College
	3 = Bachelor
	4 = Graduate
	7 = NAP
	8 = Don't Know
	9 = NA
Padeg	Father's highest degree of education
	0 = Less than High School
	1 = High School
	2 = Junior College
	3 = Bachelor
	4 = Graduate
	7 = NAP
	8 = Don't Know
	9 = NA
Madeg	Mother's highest degree of education
	0 = Less than High School
	1 = High School
	2 = Junior College
	3 = Bachelor
	4 = Graduate
	7 = NAP
	8 = Don't Know
	9 = NA
Sex	Respondent's sex (1 = M, 2 = F)
Race	Respondent's race (1 = White, 2 = Black, 3 = Other)
Region	Geographic region of respondent
	0 = NA
	1 = New England
	2 = Middle Atlantic
	3 = East North Central
	4 = West North Central
	5 = South Atlantic
	6 = East South Central
	7 = West South Central
	8 = Mountain
	9 = Pacific
Polviews	Think of self as liberal or conservative?
	0 = NAP
	1 = Extremely Liberal

          2 = Liberal
          3 = Slightly Liberal
          4 = Moderate
          5 = Slightly Conservative
          6 = Conservative
          7 = Extremely Conservative
          8 = Don't Know
          9 = NA

Grass   Should marijuana be made legal?
          0 = NAP
          1 = Legal
          2 = Not Legal
          8 = Don't Know
          9 = NA

Relig   Respondent's religious preference
          1 = Protestant
          2 = Catholic
          3 = Jewish
          4 = None
          5 = Other
          8 = Don't Know
          9 = NA

Tvhours   Hours per day spent watching TV (in hours, except as noted)
          -1 = NAP
          98 = Don't Know
          99 = NA

Partners   How many sex partners respondent has had in past year
          -1 = NAP
          0 = No partners
          1 = 1 partner
          2 = 2 partners
          3 = 3 partners
          4 = 4 partners
          5 = 5–10 partners
          6 = 11–20 partners
          7 = 21–100 partners
          8 = More than 100 partners
          9 = 1 or more, Don't know number
          95 = Several
          98 = Don't Know
          99 = NA

Sexfreq   Frequency of sex during the last year
          -1 = NAP
          0 = Not at all
          1 = Once or twice
          2 = Once a month

3 = 2–3 times a month
4 = Weekly
5 = 2–3 per week
6 = 4+ per week
8 = Don't Know
9 = NA

🖳 All of the abortion questions below are coded identically with Abany.

Abany   Should women be able to get an abortion for any reason?
                   0 = NAP
                   1 = Yes
                   2 = No
                   8 = Don't Know
                   9 = NA
Abdefect   Abortion if chance of serious defect?
Abhlth   Abortion if woman's health seriously endangered?
Abnomore   Abortion if married but wants no more children?
Abpoor   Low income- can't afford more children?
Abrape   Pregnant as a result of rape?
Absingle   Not married?
Attend   How often respondent attends religious services
                   0 = Never
                   1 = Less than once a year
                   2 = Once a year
                   3 = Several times a year
                   4 = Once a month
                   5 = 2–3 times a month
                   6 = Nearly every week
                   7 = Every week
                   8 = More than once a week
                   9 = Don't Know, NA
Childs   Number of Children (number, except as noted)
                   8 = Eight or more
                   9 = NA
Class   Subjective class identification
                   0 = NAP
                   1 = Lower class
                   2 = Working class
                   3 = Middle class
                   4 = Upper class
                   5 = No class
                   8 = Don't Know
                   9 = NA
Divorce   Ever been divorced or separated?
                   0 = NAP

           1 = Yes
           2 = No
           8 = Don't Know
           9 = NA

Evpaidsx   Ever paid for or received payment for sex since turning 18?
           0 = NAP
           1 = Yes
           2 = No
           8 = Don't Know
           9 = NA

Evstray   Ever had sex with someone other than spouse while married?
           0 = NAP
           1 = Yes
           2 = No
           3 = Never married
           8 = Don't Know
           9 = NA

Hapmar   Happiness of marriage
           0 = NAP
           1 = Very happy
           2 = Pretty happy
           3 = Not too happy
           8 = Don't Know
           9 = NA

Homosex   Are homosexual sex relations wrong?
           0 = NAP
           1 = Always wrong
           2 = Almost always wrong
           3 = Sometimes wrong
           4 = Not wrong at all
           5 = Other
           8 = Don't Know
           9 = NA

Pray   How often does respondent pray?
           0 = NAP
           1 = Several times a day
           2 = Once a day
           3 = Several times a week
           4 = Once a week
           5 = Less than once a week
           6 = Never
           8 = Don't Know
           9 = NA

Racpres   Would respondent vote for a black president?
           0 = NAP
           1 = Yes

                        2 = No
                        8 = Don't Know
                        9 = NA

> 🖥 All of the suicide questions below are coded the same as Suicide1.

Suicide1   Suicide okay if suffering from an incurable disease?
                        0 = NAP
                        1 = Yes
                        2 = No
                        8 = Don't Know
                        9 = NA
Suicide2   Suicide okay if bankrupt?
Suicide3   Suicide okay if one dishonors family?
Suicide4   Suicide okay if tired of living?
Xmarsex   Okay to have sex with person other than spouse?
                        0 = NAP
                        1 = Always wrong
                        2 = Almost always wrong
                        3 = Sometimes wrong
                        4 = Not wrong at all
                        5 = Other
                        8 = Don't Know
                        9 = NA
Income   Total family income
                        0 = NAP
                        1 = Less than $1000
                        2 = $1000–2999
                        3 = $3000–3999
                        4 = $4000–4999
                        5 = $5000–5999
                        6 = $6000–6999
                        7 = $7000–7999
                        8 = $8000–9999
                        9 = $10000–14999
                        10 = $15000–19999
                        11 = $20000–24999
                        12 = $25000 or more
                        13 = Refused
                        98 = Don't Know
                        99 = NA
Rincome   Respondent's income
                        0 = NAP
                        1 = Less than $1000
                        2 = $1000–2999
                        3 = $3000–3999

```
                    4 = $4000–4999
                    5 = $5000–5999
                    6 = $6000–6999
                    7 = $7000–7999
                    8 = $8000–9999
                    9 = $10000–14999
                    10 = $15000–19999
                    11 = $20000–24999
                    12 = $25000 or more
                    13 = Refused
                    98 = Don't Know
                    99 = NA
Marnomar   Bad marriage better than none at all?
                    0 = NAP
                    1 = Strongly agree
                    2 = Agree
                    3 = Neither agree nor disagree
                    4 = Disagree
                    5 = Strongly disagree
                    8 = Can't choose
                    9 = NA
Marhappy   Married people happier than unmarried?
                    0 = NAP
                    1 = Strongly agree
                    2 = Agree
                    3 = Neither agree nor disagree
                    4 = Disagree
                    5 = Strongly disagree
                    8 = Can't choose
                    9 = NA
```

> 🖥 All of the "Who in the household" questions below are coded identically with Laundry.

```
Laundry    Who in the household does the laundry?
                    0 = NAP
                    1 = Always the woman
                    2 = Usually the woman
                    3 = About equal or both together
                    4 = Usually the man
                    5 = Always the man
                    6 = Is done by a third person
                    7 = Not married
                    8 = Can't choose
                    9 = NA
Repairs    Who in the household does small repairs?
```

Caresick   Who cares for the sick in the family?
Shopfood   Who shops for the groceries?
Dinner   Who plans the meals?
Socbar   How often respondent spends the evening at a bar
                                                     −1 = NAP
                                                   1 = Almost daily
                                                   2 = Several times a week
                                                   3 = Several times a month
                                                   4 = Once a month
                                                   5 = Several times a year
                                                   6 = Once a year
                                                   7 = Never
                                                   8 = Don't Know
                                                   9 = NA
Fear   Afraid to walk at night in neighborhood
                                                   0 = NAP
                                                 1 = Yes
                                                 2 = No
                                                 8 = Don't Know
                                                 9 = NA
Age4cat   NTILES of AGE (quartiles of Age)

HAIRCUT.SAV

Data randomly selected from the file **STUDENT.SAV**, recording the prices paid for most recent haircut. n = 60. Source: Author.

Var.	Description
Haircut	Price paid for most recent professional haircut
Sex	Gender of the student (M/F)
Region	Home region of the student (Rural, Suburban, Urban)

HELPING.SAV

Results of a student project investigating helping behavior. A "victim" drops books on staircase in an academic building, and helping behavior of the subject is recorded. n = 45. Source: Tara O'Brien and Benjamin White. Used with permission.

Var.	Description
Subject	Subject number
Gendvict	Gender of victim (1 = Male, 2 = Female)
Gendsubj	Gender of subject (1 = Male, 2 = Female)
Helping	Does subject help? (1 = Help, 2 = No help)

IMPEACH.SAV

Results of the U.S. Senate votes in the impeachment trial of President Clinton. $n = 100$. Source: Professor Alan Reifman, Texas Tech University. Available from the *Journal of Statistics Education*. http://www.amstat.org/publications/jse. Used with permission.

Var.	Description
Name	Senator's name
Perj_1	Vote on Article I, Perjury (0 = Not guilty, 1 = Guilty)
Obstr_2	Vote on Article II, Obstruction of Justice (0 = Not guilty, 1 = Guilty)
Numguilt	Number of guilty votes cast
Party	Senator's party (0 = Democrat, 1 = Republican)
Conserv	Rating by American Conservative Union
Clint96	Percent of home state vote for Clinton in 1996
Seat_up	Year in which senator's term ends
Fresh	Is senator in first term? (0 = No, 1 = Yes)

INFANT.SAV

Experimental data on infant cognition. Infants are shown various images, and researcher records time spent looking at the image. $n = 88$. Source: Professor Lincoln Craton, Stonehill College. Used with permission.

Var.	Description
Subject	Subject number
Age	Infant age (month & days)
Sex	Subject's sex (1 = Female, 2 = Male)
Order	Stimulus order (1 = Broken/Complete, 2 = Complete/Broken)
Totbroke	Total broken looking time (seconds)
Totcompl	Total complete looking time (seconds)

LABOR.SAV

Monthly data on employment measures from the U.S. economy, Jan. 1948–Mar. 1996. $n = 614$. Source: *Survey of Current Business*.

Var.	Description
Date	Month and year
A0M441	Civilian labor force (thous.)
A0M442	Civilian employment (thous.)
A0M451	Labor force participation rate, males 20 & over (pct.)
A0M452	Labor force participation rate, females 20 & over (pct.)
A0M453	Labor force participation rate, 16–19 years of age (pct.)
A0M001	Average weekly hours, mfg. (hours)

A0M021	Average weekly overtime hours, mfg. (hours)
A0M005	Average weekly initial claims, unemploy. insurance (thous.)
A0M046	Index of help-wanted ads in newspapers (1987 = 100)
A0M060	Ratio, help-wanted advertising to number unemployed
A0M048	Employee hours in nonagricultural establishments (bil. hours)
A0M042	Persons engaged in nonagricultural activities (thous.)
A0M041	Employees on nonagricultural payrolls (thous.)
D1M963	Private nonagricultural employment, 1-mo. diffusion index (%)
D6M963	Private nonagricultural employment, 6-mo. diffusion index (%)
A0M040	Nonagricultural employees, goods-producing industries (thous.)
A0M090	Ratio, civilian employment to working-age pop.(%)
A0M037	Number of persons unemployed (thous.)
A0M043	Civilian unemployment rate (pct.)
A0M045	Average weekly insured unemployment rate (pct.)
A0M091	Average duration of unemployment in weeks (weeks)
A0M044	Unemployment rate, 15 weeks and over (pct.)

LONDON1.SAV

Selected hourly measurements of carbon monoxide concentrations in the air in West London, 1996. All measurements are parts per million (ppm). $n = 325$. Source: National Environmental Technology Centre. Data available at:

http://www.aeat.co.uk/netcen/aqarchive/data/autodata/1996/wl_co.csv

Var.	Description
Date	Day of year
CO1AM	ppm, CO for the hour ending 1AM GMT
CO5AM	ppm, CO for the hour ending 5AM GMT
CO9AM	ppm, CO for the hour ending 9AM GMT
CO1PM	ppm, CO for the hour ending 1PM GMT
CO5PM	ppm, CO for the hour ending 5PM GMT
CO9PM	ppm, CO for the hour ending 9PM GMT

LONDON2.SAV

Hourly measurements of carbon monoxide concentrations in the air in West London, 1996. All measurements are parts per million (ppm). $n = 339$. Source: National Environmental Technology Centre. Data available via:

http://www.aeat.co.uk/netcen/aqarchive/data/autodata/1996/wl_co.csv

Var.	Description
Date	Day of year
CO1AM	ppm, CO for the hour ending 1AM GMT

CO2AM	ppm, CO for the hour ending 2AM GMT
CO3AM	ppm, CO for the hour ending 3AM GMT
CO4AM	ppm, CO for the hour ending 4AM GMT
CO5AM	ppm, CO for the hour ending 5AM GMT
CO6AM	ppm, CO for the hour ending 6AM GMT
CO7AM	ppm, CO for the hour ending 7AM GMT
CO8AM	ppm, CO for the hour ending 8AM GMT
CO9AM	ppm, CO for the hour ending 9AM GMT
CO10AM	ppm, CO for the hour ending 10AM GMT
CO11AM	ppm, CO for the hour ending 11AM GMT
CO12NOON	ppm, CO for the hour ending 12 NOON GMT
CO1PM	ppm, CO for the hour ending 1PM GMT
CO2PM	ppm, CO for the hour ending 2PM GMT
CO3PM	ppm, CO for the hour ending 3PM GMT
CO4PM	ppm, CO for the hour ending 4PM GMT
CO5PM	ppm, CO for the hour ending 5PM GMT
CO6PM	ppm, CO for the hour ending 6PM GMT
CO7PM	ppm, CO for the hour ending 7PM GMT
CO8PM	ppm, CO for the hour ending 8PM GMT
CO9PM	ppm, CO for the hour ending 9PM GMT
CO10PM	ppm, CO for the hour ending 10PM GMT
CO11PM	ppm, CO for the hour ending 11PM GMT
CO12MID	ppm, CO for the hour ending 12 MIDNIGHT GMT

MARATHON.SAV

Finishing times and rankings for the Wheelchair division of the 1996 Boston Marathon. $n = 81$. Source: Boston Athletic Association and *Boston Globe*. http://www.boston.com/sports/marathon

Var.	Description
Rank	Order of finish
Name	Name of racer
City	Home city or town of racer
State	Home state or province of racer
Country	Three-letter country code
Minutes	Finish time, in minutes

MENDEL.SAV

Summary results of one genetics experiment conducted by Gregor Mendel. Tally of observed and frequencies and expected relative frequencies of pea texture and color. $n = 4$. Source: Heinz Kohler, *Statistics for Business and Economics*, 3rd ed. (New York: HarperCollins, 1994), p. 459.

Var.	Description
Type	Identification of color and texture
Observed	Frequency observed
Expected	Expected frequency

MFT.SAV

This worksheet holds scores of students on a Major Field Test, as well as their GPAs and SAT verbal and math scores. $n = 137$. Source: Prof. Roger Denome, Stonehill College. Used with permission.

Var.	Description
TOTAL	Total score on the Major Field Test
SUB1	Score on Part 1
SUB2	Score on Part 2
SUB3	Score on Part 3
SUB4	Score on Part 4
GPA	Student's college GPA at time of exam
Verb	Verbal SAT score
Math	Math SAT score
GPAQ	Quartile in which student's GPA falls in sample
VerbQ	Quartile in which student's Verbal SAT falls in sample
MathQ	Quartile in which student's Math SAT falls in sample

MILGRAM.SAV

Results of four of Professor Stanley Milgram's famous experiments in obedience. $n = 160$. Source: Stanley Milgram, *Obedience to Authority* (New York: Harper, 1975), p. 35.

Var.	Description
Exp	Experiment Number
Volts	Volts administered

NIELSEN.SAV

A. C. Nielsen television ratings for the top 20 shows, as measured during the week of September 14, 1997. $n = 20$. Source: A. C. Nielsen Co.

Var.	Description
Rank	Ranking of the show (1 through 20)
Show	Title of the program
Network	Code identifying broadcast network
Rating	Rating score for the program that week

NORMAL.SAV

Artificial data to illustrate features of normal distributions. $n = 100$. Source: Authors.

Var.	Description
X	Values of a continuous random variable
CN01	Cumulative Normal densities for $X \sim N(0,1)$
CN11	Cumulative Normal densities for $X \sim N(1,1)$
CN03	Cumulative Normal densities for $X \sim N(0,3)$
N01	Normal densities for $X \sim N(0,1)$
N11	Normal densities for $X \sim N(1,1)$
N03	Normal densities for $X \sim N(0,3)$
Value	User-entered value
Cumprob	Cumulative probability corresponding to Value
Hundred	A sequence of values from 1 through 100
Binomial	Computed binomial probabilities

OUTPUT.SAV

Monthly data on output, production and capacity utilization measures from the U.S. economy, Jan. 1948–Mar. 1996. $n = 616$. Source: *Survey of Current Business*.

Var.	Description
Date	Month and year
A0M047	Index of industrial production (1987 = 100)
A0M073	Indust. production, durable goods manufacturers (1987 = 100)
A0M074	Indust. production, nondurable manuf. (1987 = 100)
A0M075	Industrial production, consumer goods (1987 = 100)
A0M124	Capacity utilization rate, total industry (pct.)
A0M082	Capacity utilization rate, manufacturing (pct.)

PAWORLD.SAV

The Penn World Table (Mark 5.6) was constructed by Robert Summers and Alan Heston of the University of Pennsylvania for an article in the May 1991 *Quarterly Journal of Economics*. The main dataset is massive, containing demographic and economic data about virtually every country in the world from 1950 to 1992. This dataset represents selected variables and a stratified random sample of 42 countries from around the world, for the period 1960–1992. $n = 1386$. Source: http://cansim.epas.utoronto.ca:5680/pwt/. Used with permission of Professor Heston.
NOTE: Countries are sorted alphabetically within continent.

Var.	Description
ID	Numeric Country code
COUNTRY	Name of Country
YEAR	Observation year (1960–1992)
POP	Population (thousands)
RGDPCH	Per capita Real GDP; Chain Index, 1985 international prices
C	Real Consumption % of GDP
I	Real Investment % of GDP
G	Real Government expenditures % of GDP
Y	Real per capita GDP relative to U.S. (%;U.S. = 100)
CGDP	Real GDP per capita, current international prices
XR	Exchange rate with U.S. dollar
RGDPEA	Real GDP per equivalent adult
RGDPW	Real GDP per worker
OPEN	"Openness" = (Exports + Imports)/Nominal GDP

PENNIES.SAV

Students in a class each flip 10 coins repeatedly until they have done so about 30 times. They record the number of heads in each of the repetitions. $n = 56$. Source: Professor Roger Denome, Stonehill College. Used with permission.

Var.	Description
Heads00	Number of times (out of 30) student observed 0 heads
Heads01	Number of times (out of 30) student observed 1 head
Heads02	Number of times (out of 30) student observed 2 heads
Heads03	Number of times (out of 30) student observed 3 heads
Heads04	Number of times (out of 30) student observed 4 heads
Heads05	Number of times (out of 30) student observed 5 heads
Heads06	Number of times (out of 30) student observed 6 heads
Heads07	Number of times (out of 30) student observed 7 heads
Heads08	Number of times (out of 30) student observed 8 heads
Heads09	Number of times (out of 30) student observed 9 heads
Heads10	Number of times (out of 30) student observed 10 heads

PHERESIS.SAV

Quality control data on blood platelet pheresis donations. Each donation is analyzed for volume of platelets and white blood cells, as well as equipment used to collect donation. $n = 294$. Source: Dr. Mark Popovsky, Blood Services, Northeast Region, American Red Cross. Used with permission.

Var.	Description
Volume	Total volume of donation
Wbc	White blood count per product
Wbn	Donation code number
Ph	pH level
Machine	Machine
Lr	Left-Right split (0 = Left, 1 = Right)
Platelet	Platelet yield
Month	Month
Maccode	Machine code
	1 = Fenwal
	2 = Fenwal Split
	3 = Cobe
	4 = Cobe Split
	5 = Amicus
	6 = Amicus Split
Sample	Sample number
Lowyield	Low platelet yield indicator
Highwbc	High white blood count indicator
Lowph	Low pH indicator
Highph	High pH indicator
Flag	Problem flag

PHYSIQUE.SAV

Results of a student experiment on social physique anxiety among female college students. Subjects were administered a social physique anxiety scale instrument, as well as a situational comfort level scale instrument. $n = 18$. Source: Stephanie Duggan and Erin Ruell. Used with permission.

Var.	Description
Subject	Subject number
Cond	Condition
Physique	Social Physique Anxiety score
SPAlevel	Social Physique Anxiety level
	1.00 = Low SPA
	2.00 = High SPA
Anxsit	Anxiety Invoking Situation score
Total	Total situational comfort score

SALEM.SAV

Taxes paid, political factions, and status during the witchcraft trials of people living in the Salem Village parish, 1690–1692. $n = 100$.

Source: Paul Boyer and Nissenbaum, Stephen, *Salem Village Witchcraft: A Documentary Record of Local Conflict in Colonial New England.* (Boston: Northeastern University Press, 1993). Used with permission.

Var.	Description
Last	Last name
First	First name
Tax	Amount of tax paid, in pounds, 1689–90
ProParri	(0–1) indicator variable identifying persons who supported Rev. Parris in 1695 records (1 = supporter)
Accuser	(0–1) indicator variable identifying accusers and their families (1 = accuser)
Defend	(0–1) indicator variable identifying accused witches and their defenders (1 = defender)

SLAVDIET.MTW

Per capita food consumption of slaves in 1860 compared with the per capita food consumption of the entire population, 1879. $n = 12$. Source: Robert William Fogel and Engerman, Stanley L., *Time on the Cross: Evidence and Methods—A Supplement.* (Boston: Little, Brown and Company, 1974).

Var.	Description
Food	Food product
Type	Food group (e.g., meat, grain, dairy, etc.)
Slavlb	Per capita lbs. consumed by slaves in 1860
Slavcal	Per capita calories per day for slaves in 1860
Poplb	Per capita lbs. consumed by general population, 1879
Popcal	Per capita calories per day, general population, 1879

SLEEP.SAV

Data describing sleep habits, size, and other attributes of mammals. $n = 62$. Source: Allison, T. and Cicchetti, D., "Sleep in Mammals: Ecological and Constitutional Correlates," *Science*, v. 194, (Nov. 12, 1976, pp. 732–734.) Used with authors' permission. Data from http://lib.stat.cmu.edu/datasets/sleep, contributed by Prof. Roger Johnson, South Dakota School of Mines and Technology.

Var.	Description
Species	Name of mammalian species
Weight	Body weight, kg.
Brain	Brain weight, grams

Sleepnon	Nondreaming sleep (hrs/day)
Sleepdr	Dreaming sleep (hrs/day)
Sleep	Total sleep (hrs/day)
LifeSpan	Maximum life span (years)
Gestat	Gestation time (days)
Predat	Predation index (1–5, from least to most likely to be preyed upon
Exposure	Sleep exposure index (1–5; 1 = sleeps in well-protected den, 5 = highly exposed)
Danger	Overall danger index (1–5: least to most)

SPINNER.SAV

Data file for simulation of repeated spins of a game spinner. $n = 1000$. Source: Authors.

Var.	Description
Spin	Sequential list of values 1 through 1000
Quadrant	Result of simulation (initially empty cells)

STATES.SAV

Data concerning population, income, and transportation in the 50 states of the U.S., plus the District of Columbia. $n = 51$. Source: Highway Statistics On-Line, 1990.

Var.	Description
State	Name of State
Pay93	Mean wages in 1993 (current dollars)
Pay94	Mean wages in 1994 (current dollars)
Chg9394	% change in wages, 1993 to 1994
Pop	Population, 1994
Area	Land area of the state, square miles
Density	Population per square mile
Ins92	Mean auto insurance premium, 1992
Ins93	Mean auto insurance premium, 1993
Ins94	Mean auto insurance premium, 1994
CarsIns	Number of cars insured, 1994
Regist	Number of cars registered, 1994
RdMiles	Number of miles of road in the state
Mileage	Avg. number of miles driven by drivers in the state
FIA	Fatal Injury Accidents
AccFat	Number of fatalities in auto accidents
BAC	Blood Alcohol Content threshold
MaleDr	Number of male drivers licensed, 1994
FemDr	Number of female drivers licensed, 1994

| TotDriv | Total number of licensed drivers, 1994 |
| RateFat | Traffic fatalities per 100,000 population |

STUDENT.SAV

This file contains results of a first-day-of-class survey of Business Statistics students at Stonehill College. All students in the sample are full-time day students. $n = 219$. Source: Author.

Var.	Description
ID	Identifier Code
Gender	F = Female, M = Male
Ht	Height, in inches
Wt	Weight, in pounds
DOW	Day of Week on which your birthday falls this year
Left	Hand you write with (0 = Right, 1 = Left)
Eyes	Eye Color (Blue, Brown, Green, Hazel)
Maj	Major field of study (ACC = accounting, FIN = finance, MGT = management, MKT = marketing, OTH = other)
Res	Resident (R) or Commuter Student (C)
WorkHr	Hours worked per week at a paid job
GPA	Current cumulative GPA in college
OwnCar	Car ownership (Y/N)
Home	Miles between your home and school (est.)
Region	Is your hometown rural (R), suburban (S), or urban (U)
Drive	"How do you rate yourself as a driver?" (1=Below Average, 2=Average, 3= Above Average)
Belt	Frequency of seat belt usage (Never, Sometimes, Usually, Always)
Acc	Number of auto accidents involved in within past 2 yrs.
Sibling	Number of siblings
Cigs	Smoked a cigarette in past month? (1 = yes)
Haircut	Price paid for most recent professional haircut
Dog	Own a dog?
Travel	Ever traveled outside of U.S.A?
Zap	Personally know someone hit by lightning (1 = yes)
Beers	Number of beers consumed on Labor Day
Female	Sex (1 = Female, 0 = Male)
WorkCat	How many hours per week do you work at a paid job? 0 = none (0 hrs.) 1 = some (1–19 hrs.) 2 = many (20–99 hrs.)

SWIMMER.SAV

This file contains individual-event race times for a high school swim team. Each swimmer's time was recorded in two "heats" (trials) of each event in which he or she competed. Times are in seconds. Each observation represents one swimmer in one heat of one event. $n = 272$. Source: Brian Carver. Used with permission.

Var.	Description
Swimmer	ID code for each swimmer
Gender	Gender (F/M)
Heat	First or second heat (1/2)
Event	Identifier of swimming event timed (length in Meters plus event code: Freestyle, Breast, Back)
EventRep	Combination of Heat and Event
Time	Recorded time to complete the event

SWIMMER2.SAV

This file contains individual-event race times for a high school swim team. Each swimmer's time was recorded in two "heats" (trials) of each event in which he or she competed. Times are in seconds. Each row contains all results for each of 72 swimmers. $n = 72$. Source: Brian Carver. Used with permission.

Var.	Description
Swimmer	ID code for each swimmer
Gender	Gender (F/M)
Events	Number of different events recorded for the swimmer
Num50	Number of 50-meter events for this swimmer
Fr10001	Time in 100-meter freestyle (1st heat)
Fr10002	Time in 100-meter freestyle (2nd heat)
Fr20001	Time in 200-meter freestyle (1st heat)
Fr20002	Time in 200-meter freestyle (2nd heat)
Bk5001	Time in 50-meter backstroke (1st heat)
Bk5002	Time in 50-meter backstroke (2nd heat)
Br5001	Time in 50-meter breaststroke (1st heat)
Br5002	Time in 50-meter breaststroke (2nd heat)
Fr5001	Time in 50-meter freestyle (1st heat)
Fr5002	Time in 50-meter freestyle (2nd heat)

TRACK.SAV

Times for selected NCAA Women running the 3000-meter indoors and outdoors. $n = 31$. Source: www.ncaaschampionships.com/sports/.

Var.	Description
Athlete	Athlete's name
YR	Year in college
School	College of athlete
Timein	Time in indoor 3000 meters
Datein	Date of indoor time
Timeout	Time in outdoor 3000 meters
Dateout	Date of outdoor time

US.SAV

Time series data about the U.S. economy during the period from 1965–1996. $n = 32$. Sources: *Statistical Abstract of the United States, Economic Report of the President*, various years.

Var.	Description
Yr	Observation Year
Pop	Population of the U.S. for the year (000s)
Employ	Aggregate Civilian Employment (000s)
Unemprt	Unemployment Rate (%)
GNP	Gross National Product (billions of current $)
GDP	Gross Domestic Product (billions of current $)
PersCon	Aggregate Personal Consumption (billions)
PersInc	Aggregate Personal Income (billions)
PersSav	Aggregate Personal Savings (billions)
DefGDP	GDP Price Deflator (1987 = 100)
DefPC	Personal Consumption/Income Deflator (1987 = 100)
CPI	Consumer Price Index (1982–84 = 100)
M1	Money supply (billions)
DOW	Dow-Jones 30 Industrials Stock Avg.
Starts	Housing Starts (000s)
Sellprc	Median selling price of a new home (current $)
ValNH	Value of new housing put in place (current mil. $)
NHMort	New home mortgage interest rate
PPIConst	Produce Price Index for construction materials
Cars	Cars in use (millions)
MortDebt	Aggregate mortgage debt (billions, current)
Exports	Total exports of goods and services (bil., current)
Imports	Total imports of goods and services (bil., current)
FedRecpt	Total Federal receipts (billions, current $)
FedOut	Total Federal outlays (billions, current $)
FedInt	Interest paid on Federal debt (billions, current)

UTILITY.SAV

Time series data about household usage of natural gas and electricity over a period of years in the author's home. $n = 81$. Source: R. Carver, "What Does it Take to Heat a New Room?" *Journal of Statistics Education* v. 6, no. 1 (1998). Used with permission. Available at http://www.amstat.org/publications/jse.

Var.	Description
Month	Month and year of observation
Days	Days in the month
MeanTemp	Mean temperature in Boston for month
GaspDay	Mean number of "therms" of natural gas consumed
Therms	Total therms used during month
GasDays	Number of billing days in month (gas)
KWH	Kilowatt-hours of electricity consumed
KWHpDay	Mean Kilowatt-hours of electricity used per day
ElecDays	Number of billing days in month (electric)
Est	Electricity bill based on actual reading or estimate · (0 = actual, 1 = estimate)
HDD	Heating degree-days in the month[2]
CDD	Cooling degree days
NewRoom	Dummy variable indicating when the house was enlarged by one room.

WATER.SAV

Data concerning freshwater consumption in 221 water regions throughout the United States, for the years 1985 (columns 2 through 17) and 1990 (columns 18 through 33). All consumption figures are in millions of gallons per day, unless otherwise noted. $n = 221$. Source: U.S. Geological Survey. http://water.usgs.gov/public/watuse/wudata.html

Var.	Description
Area	Region identifier code
Poptot85	Total population of area, thousands
Pswtfr85	Total freshwater withdrawals
Pspcap85	Per capita water use, gallons per day
Cocuse85	Commercial consumptive use

---

[2] A "degree-day" equals the sum of daily mean temperature deviations from 65° F. For heating degree days, only days below 65° F are counted. For cooling degree days, only days warmer than 65° F are counted.

Docuse85	Domestic consumptive use
Incufr85	Industrial freshwater consumptive use
Ptcufr85	Thermoelectric power freshwater consumptive use
Pfcufr85	Thermoelectric power (fossil fuel) freshwater consumptive use
Pgcufr85	Thermoelectric power (geothermal) freshwater consumptive use
Pncufr85	Thermoelectric power (nuclear) freshwater consumptive use
Micufr85	Mining freshwater consumptive use
Lvcuse85	Livestock freshwater consumptive use
Irconv85	Irrigation conveyance losses
Ircuse85	Irrigation freshwater consumptive use
Tofrto85	Total freshwater use (all kinds combined)
Tocufr85	Total freshwater consumptive use
Poptot90	Total population of area, thousands
Pswtfr90	Total freshwater withdrawals
Pspcap90	Per capita water use, gallons per day
Cocuse90	Commercial consumptive use
Docuse90	Domestic consumptive use
Incufr90	Industrial freshwater consumptive use
Ptcufr90	Thermoelectric power freshwater consumptive use
Pfcufr90	Thermoelectric power (fossil fuel) freshwater consumptive use
Pgcufr90	Thermoelectric power (geothermal) freshwater consumptive use
Pncufr90	Thermoelectric power (nuclear) freshwater consumptive use
Micufr90	Mining freshwater consumptive use
Lvcuse90	Livestock freshwater consumptive use
Irconv90	Irrigation conveyance losses
Ircuse90	Irrigation freshwater consumptive use
Tofrto90	Total freshwater use (all kinds combined)
Tocufr90	Total freshwater consumptive use
Pctcu85	Consumptive use at % of total use, 1985
Frchgpc	% change in freshwater use, 1985 to 1990
Frchg	Change in freshwater use, 1985 to 1990

WEB.SAV

Results of 20 sets of 20 trials using the Random Yahoo! Link. "Problem" defined as encountering an error message or message indicating that the referenced site had moved. $n = 20$. Source: Yahoo!®. Use this bookmark to activate the Random Yahoo! Link: http://random.yahoo.com/bin/ryl

Var.	Description
Sample	Sample number (1–20)
N	Sample size (equals 20 in all samples)
Problems	Number of problems encountered in n repetitions.

WORLD90.SAV

This file is extracted from the Penn World Tables dataset described above. All data refer only to the year 1990. $n = 42$. Source: See **PAWORLD.SAV**, (page 297).

Var.	Description
ID	Numeric Country code
COUNTRY	Name of Country
POP	Population in 000s
RGDPCH	Per capita Real GDP, using a Chain Index, 1985 international prices
C	Real Consumption % of GDP
I	Real Investment % of GDP
G	Real Government expenditures % of GDP
Y	Real per capita GDP relative to U.S. (%; U.S. = 100)
CGDP	Real GDP per capita, current international prices
XR	Exchange rate with U.S. dollar
RGDPEA	Real GDP per equivalent adult
RGDPW	Real GDP per worker
RGDP88	Real GDP per capita, 1988

XSQUARE.SAV

Data to illustrate a deterministic quadratic relationship. $n = 20$. Source: Authors.

Var.	Description
X	Sequence of values from 1 through 20
Xsquare	The square of x
Y	A quadratic function of x

Appendix B

Working with Files

Objectives

This Appendix explains several common types of files that SPSS supports and uses. Though you may not use each kind of file, it will be helpful to understand the distinctions among them. Each file type is identified by a three-character extension (such as .SAV or .SPO) to help distinguish them. For those just getting started with statistics and SPSS, the most useful file types are these:

Extension	File Type
.SAV	SPSS Data file
.SPO	SPSS Viewer Document
.POR	SPSS Portable Data file
.SYS	SPSS/PC+ Data file
.SPS	SPSS Syntax (i.e. macro)

The following sections review these types of files, and explain their use. In addition, there is a section which illustrates how you can convert data from a spreadsheet into a SPSS worksheet.

Data Files

Throughout this manual, you have read data from SPSS data files. These files have the extension .SAV, and the early exercises explain how to open and save such files. These files just contain raw data (numeric, text, or date/time), as well as variable and value labels.

Generally, when you enter data into the Data Editor, the default settings of column format (data type, column width, and so on) are acceptable. Should you wish to customize some of these elements, you'll find relevant commands on the Variable View tab, which is explained in Session 1.

In earlier versions of SPSS, the information embedded in a data file was slightly different. This is one reason the **File ➤ Save As...** dialog box lists several SPSS formats (among others) for saving data files, as shown here:

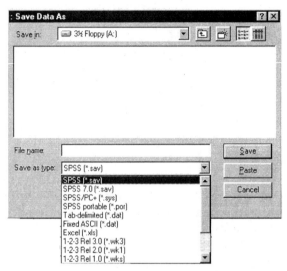

Also note that data in a worksheet can be saved as a SPSS Portable file, or in one of several popular spreadsheet or database formats. The latter options are discussed later. The Portable format creates a data file that can be read in SPSS running under other operating systems (e.g. UNIX or Macintosh). If you need to save a worksheet for other SPSS users, but aren't sure which version of SPSS they run, this is the safe choice of file formats.

Note that when you save a data file, you are *only* saving the data from a given session. If you also want to save the results of analysis, you must save the Viewer document.

Viewer Document Files

After doing analysis with SPSS, you may want to save a record of the work you've done, particularly if you need to complete it at a later time. That is the point of the Viewer documents. These files are "transcripts" of the outputs you have generated during a working session. A Viewer document contains everything that you see in the Viewer window. Both the outline and content panes are saved.

To save a Viewer document, first be sure that you are in the Viewer window. Then, give this command:

- **File ➤ Save...** This will bring up the Save As dialog box; give your document a name, and click **OK**. Note that, by default, SPSS assigns the .SPO suffix to the file name.

Converting Other Data Files into SPSS Data Files

Often one might have data stored in a spreadsheet or database file, or want to analyze data downloaded from the Internet. SPSS can easily open many types of files. This section discusses two common scenarios; for other file types, you should consult the extensive Help files provided with SPSS. Though you do not have the data files illustrated here, try to follow these examples with your own files, as needed.

Excel Spreadsheets

Suppose you have some data in a Microsoft Excel spreadsheet, and wish to read it into the SPSS Data Editor. You may have created the spreadsheet at an earlier time, or downloaded it from the Internet. This example shows how to open the spreadsheet from SPSS.[1] Such a spreadsheet is shown here:

Student	Quiz1	Midterm	Quiz2	Homework	FinalExam
Appel	97	95	88	95	92
Boyd	85	90	88	85	78
Chamberlin	86	86	82	74	80
Drury	73	68	70	55	65
Elliott	64	70	72	73	78
Franklin	82	90	75	80	85
Gooding	98	96	94	98	95
Harriman	80	82	86	74	88
Ingram	100	97	100	100	95
Jenkins	78	71	76	77	75
Kennedy	72	60	68	80	77

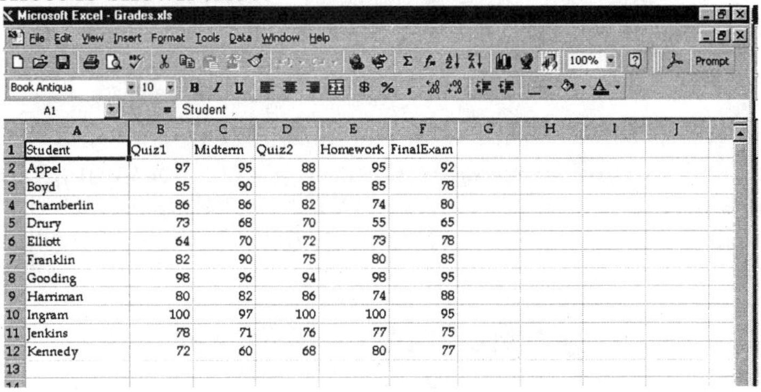

[1] What we describe here can also be accomplished by the **Database Capture** command on the File menu. We regard the latter approach to be more complex, and therefore less appropriate for new users.

First, it helps to structure the spreadsheet with variable names in the top row, and reserve each column for a variable. Though not necessary, it does simplify the task. If you are using SPSS 9.0, you must also be sure that the spreadsheet was saved as an Excel 4.0 spreadsheet. SPSS 10.0 supports more current versions.

Assuming the spreadsheet has been saved as an Excel (.XLS) file, called **Grades**, you would proceed as follows in the SPSS Data Editor:

File ➤ Open ➤ Data... In the dialog box, choose the appropriate drive and directory, and select Excel (*.xls) as the file type. You should see your file listed. Select the desired file name and click **Open**.

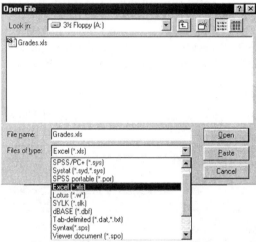

At this point, SPSS will ask if the spreadsheet contains variable names, and will ask you to specify the range of cells containing the data. This file does include variable names, so we check that box. SPSS will read all available data if we leave the Range box blank; if we wished to read only some rows, we could specify a range in standard Excel format (e.g., A1:D10).

If information in the spreadsheet is arranged as described earlier (variables in columns, and variable names in the first row), you can easily leave the Range box empty. However, if the data begin in "lower" rows of the spreadsheet, this dialog box permits you to specify where SPSS will find the data.

When you click OK, SPSS switches to the Output Viewer to display a log reflecting the fact that the data have been read into the Data Editor.

🖑 Switch back to the Data Editor, and you will find the data, as shown here. You now can analyze it and save it as a SPSS data file.

	student	quiz1	midterm	quiz2	homework	finalexa
1	Appel	97.00	95.00	88.00	95.00	9
2	Boyd	85.00	90.00	88.00	85.00	7
3	Chamberlin	86.00	86.00	82.00	74.00	8
4	Drury	73.00	68.00	70.00	55.00	6
5	Elliott	64.00	70.00	72.00	73.00	7
6	Franklin	82.00	90.00	75.00	80.00	8
7	Gooding	98.00	96.00	94.00	98.00	9
8	Harriman	80.00	82.00	86.00	74.00	8
9	Ingram	100.00	97.00	100.00	100.00	9
10	Jenkins	78.00	71.00	76.00	77.00	7
11	Kennedy	72.00	60.00	68.00	80.00	7

Data in Text Files

Much of the data available for downloading from Internet sites is in text files, sometimes referred to as ASCII format.[2] As just described, SPSS can read data from these files, but needs to be told how the data are arranged in the file.

In these files, observations appear in rows or lines of the file. Text files generally distinguish one variable from another either by following a

---

[2] ASCII stands for American Standard Code for Information Interchange. Unlike a SPSS worksheet or other spreadsheet format, an ASCII file contains no formatting information (fonts, etc.), and only contains the characters which make up the data values.

fixed spacing arrangement, or by using a separator or "delimiter" character between values. Thus, with fixed spacing, the first several rows of the student grade data might look like this:

```
Appel        97    95    88    95    92
Boyd         85    90    88    85    78
Chamberlin   86    86    82    74    80
Drury        73    68    70    55    65
```

In this arrangement, the variables occupy particular positions along the line. For instance, the student's name is within the first 11 spaces of a given line, and his or her first quiz value is in positions 12 through 14.

Alternatively, some text files don't space the data evenly, but rather allow each value to vary in length, inserting a prespecified character (often a comma or a tab) between the values, like this:

```
Appel, 97, 95, 88, 95, 92
Boyd, 85, 90, 88, 85, 78
Chamberlin, 86, 86, 82, 74, 80
Drury, 73, 68, 70, 55, 65
```

Logically, both of these lists of data contain all of the same information. To our eyes and minds, it is easy to distinguish that each list represents six variables. Though there is no single "best format" for a text file, it is important for us to correctly identify the format to SPSS, so that it can correctly import the data into the Data Editor. For this example, we'll assume the text data are stored in a file named **Grades.txt**. Within the Data Editor, give this command:

🖱 **File ➤ Open ➤ Data...** As always, select the appropriate file path, and choose Text (*.txt) next to Files of type. Choose the file from the list of available files, and click **Open**.

This will initiate the Text Import Wizard, a series of six dialog boxes that provide a number of choices about the format and organization of the data. Frequently, the default choices are already correct, and you merely need to proceed to the next dialog box. Because each text file is different, we don't show a full example here. The wizard leads you through the process quite clearly, and also offers on-line Help.

Though this discussion has not covered all possibilities, it does treat several common scenarios. By using the Help system available with your software, and patiently experimenting, you will be able to handle a wide range of data sources.

Appendix C

A First Look at SPSS 9.0

Objectives

In this session, you will learn to do the following:

- Launch and exit SPSS
- Enter quantitative and qualitative data in a data file
- Create and print a graph
- Get Help
- Save your work to a diskette

Launching SPSS

Before starting this session, you should know how to run a program within the Windows 95, Windows 98, or Windows NT operating system. All of the instructions in this manual presume basic familiarity with the Windows environment.

> Check with your instructor for specific instructions about running Windows 95/98/NT on your system. Your instructor will also tell you where to find SPSS.

Click and hold the left mouse button on the **Start** button at the lower left of your screen, and drag the cursor to select **Programs**. In the list, locate and choose **SPSS 9.0 for Windows**. Click and release the mouse button to launch the program. Because SPSS is a large program, you will have to wait a few moments before the program is ready for use.

On the next page is an image of the screen you will see when SPSS is ready. First you will see a menu dialog box listing several options; behind it is the *Data Editor* which is used to display the data that you will analyze using the program. Later you will encounter the *output Viewer window* which displays the results of your analysis. Each window has a unique purpose, to be made clear in due course. It's important at the outset to have a sense of what each window is about.

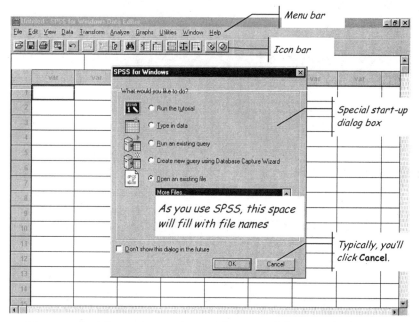

At any point in your session, only one of the windows is *selected*, meaning that mouse actions and keystrokes will affect that window alone. When you start SPSS, the Data Editor is initially selected.

Since SPSS operates upon data, we generally start by placing data into a worksheet, either from the keyboard or from a stored disk file. The Data Editor looks much like a spreadsheet. Cells may contain numbers or text, but unlike a spreadsheet, they never contain formulas. Except for the top row, which is reserved for variable names, rows are numbered consecutively. Each variable in your dataset will occupy one column of the SPSS worksheet, and each row represents one observation. For example, if you have a sample of fifty observations on two variables, your worksheet would contain two columns and fifty rows.

The menu bar across the top of the screen identifies broad categories of SPSS's features. There are two ways to issue commands in SPSS: choose commands from the menu or icon bars, or type them directly into a Syntax Editor. This book always refers you to the menus and icons. You can do no harm by clicking on a menu and reading the choices available, and you should expect to spend some time exploring your choices in this way.

Entering Data into the Data Editor

For most of the sessions in this book, you will start by accessing data already stored on a disk. For small datasets or class assignments, though, it will often make sense simply to type in the data yourself. For this session, you will transfer the data displayed below into the empty worksheet in the Data Editor.

In this first session, our goal is simple: to create a small data file, and then use SPSS to construct a graph using the data. This is typical of the tasks you will perform throughout the book.

The coach of a high school swim team runs a practice for 10 swimmers, and records their times (in seconds) on a piece of paper.[1] Each swimmer is practicing the 50-meter freestyle event, and the boys on the team assert that they did better than the girls. The coach wants to analyze these results to see what the facts are. He codes gender with a 1 for the girls and a 2 for the boys.

Swimmer	Gender	Time
Sara	1	29.34
Jason	2	30.98
Joanna	1	29.78
Donna	1	34.16
Phil	2	39.66
Hanna	1	44.38
Sam	2	34.80
Ben	2	40.71
Abby	1	37.03
Justin	2	32.81

The first step in entering the data into the Data Editor is to define three variables: Swimmer, Gender, and Time. Creating a variable

---

[1] Nearly every dataset in this book is real. For the sake of starting modestly, we have taken a minor liberty in this session. This example is actually extracted from a dataset you will use later in the book. The full dataset appears in two forms: **Swimmer** and **Swimmer2**.

requires us to name it, identify the type of data (qualitative, quantitative, number of decimal places, etc.), and assign labels to the variable and data values if we wish.

🖰 Move the cursor into the Data Editor, and position it in the left-most gray rectangle marked **Var** and double-click. You will see this dialog box; complete it by following the numbered steps, which are explained below.

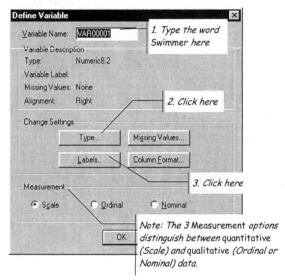

🖰 Type the variable name Swimmer as shown above.

🖰 Now, click the button marked **Type**, and you'll see this dialog box to the right. Numeric is the default variable type.

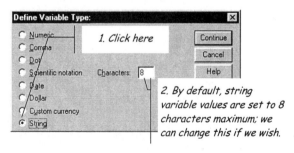

🖰 Click on the circle labeled String in the lower left corner of the dialog box. The names of the swimmers constitute a *nominal* or *categorical* variable, represented by a "string" of characters rather than a number. Click **Continue**.

Notice that Nominal is now selected as the Measurement option, because you chose String as the variable type.

🖱 Now click the **Labels** button, complete the dialog box as shown here, and click **Continue**, and then click **OK**.

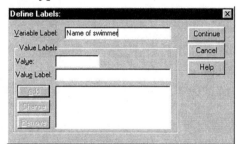

In SPSS, each variable may carry a descriptive label to help identify its meaning. Additionally, as we'll soon see, we can also label individual values of a variable.

🖱 Now let's create the Gender variable in the second column. Define the name of the variable as you did for Name. This time, though, the Type of the variable will be numeric. Complete the Define Variable Type dialog box as shown here:

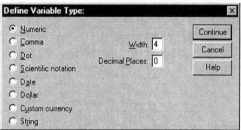

In the Define Labels dialog box, we'll create value labels:

🖱 In the box marked Variable Label, type Sex of swimmer.

🖱 In the Value Labels area of the dialog box, type 1 in the box marked Value, and Female in the Value Label box. Click **Add**.

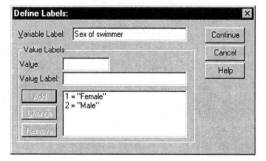

🖱 Then type 2 in Value, and Male in Value Label. Click **Add**. The dialog box should now look like this; when it does, click **Continue** and then click **OK**.

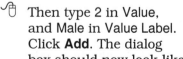

Finally, we'll create a *scale* variable in this dataset: Time. Scale variables are quantitative or numerical.

🖰 Begin as you have done twice now, by naming the third variable Time. You may leave **Type** as it is, since the default setting of 8 spaces wide with two decimal places is appropriate here.[2]

🖰 Label this variable "Practice time (secs)."

At this point, you are ready to type in the data. Follow these directions, using the data table found on page 3. If you make a mistake, just return to the cell and re-type the entry.

🖰 Position the cursor in the first cell below Swimmer, and type Sara, then press the Enter key. The cursor will move to the next cell, and you should type Jason.

🖰 After completing the names, use the mouse or arrow keys to go to the top cell under Gender, and continue.

When you are done, the Data Editor should look like this:

	name	gend	time	var	var
1	Sara	1	29.34		
2	Jason	2	30.98		
3	Joanna	1	29.78		
4	Donna	1	34.16		
5	Phil	2	39.66		
6	Hanna	1	44.38		
7	Sam	2	34.80		
8	Ben	2	40.71		
9	Abby	1	37.03		
10	Justin	2	32.81		

[2] When we create a numeric variable, we specify the maximum length of the variable and the number of decimal places. For example, the data type Numeric 8.2 refers to a number eight characters long, of which the final two places follow the decimal point: e.g., 12345.78.

Saving a Data File

It is wise to save all of your work in a disk file. SPSS distinguishes between two types of files that one might want to save. As time goes by, you might want to save your output or a data set.

At this point, we've created a data file and ought to save it on a diskette. Let's call the data file **Swim**.

> 🖳 Check with your instructor to see if you can save the data file on a hard drive or network drive in your system.

> 🖱 On the **File** menu, choose **Save As...**. In the **Save in** box, select **3½ Floppy (A:)**. Then, next to **File Name**, type swim. Click **Save**.

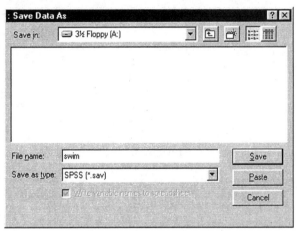

Creating a Bar Chart

With the data entered and saved, we can begin to look for an answer for the coach. We'll first use a bar chart to display the average time for the males in comparison to the females.

> 🖱 Click on **Graphs** in the menu bar, and choose **Bar...**. You will see the dialog box shown to the right.

As you can see, there are several options available. This is true for many commands; we'll typically use the default options early in this book, moving to other choices as you become more familiar with statistics and with SPSS.

Click on the button marked **Define**.

Now you should see this dialog box; complete it following the numbered steps, which are explained further below.

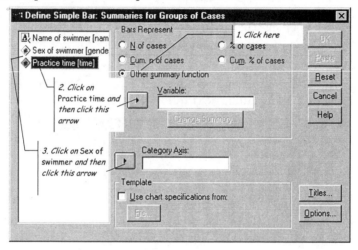

Notice that the three variables are initially listed by description and name on the left side of the dialog box. Also note that the type of variable is denoted by a symbol:

         A̹   Nominal variable (qualitative)

         ⊛   Scale variable (quantitative)

Click the radio button marked Other summary function.

In the list of variables, click on Practice time [time], and then click the arrow button next to the box marked Variable. Notice that the expression Mean(Practice time [time]) appears in the box; the chart will display the mean, or average, practice times.

Now click on Sex of swimmer, and click the arrow next to the box marked Category Axis. Then click **OK**.

You will now see a new window appear, containing a bar chart (see facing page). This is the output Viewer, and contains two panes. On the left is the *Outline pane*, which displays an outline of all of your output. The *Content pane*, on the right, contains the output itself.

Also, notice the menu bar at the top of the Viewer window. It is very similar to the menu bar in the Data Editor, with some minor differences. In general, we can perform statistical analysis from either window. Later, we'll learn some data manipulation commands that can only be given from the Data Editor.

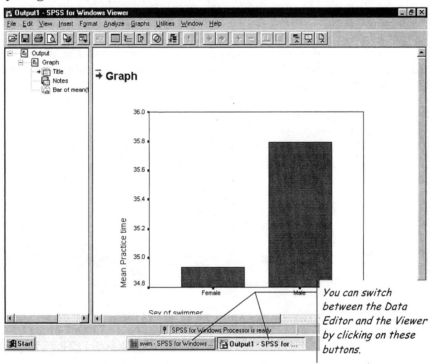

You can switch between the Data Editor and the Viewer by clicking on these buttons.

The first thing you should do is to put your name on your output. Eventually when you print the output, you will be able to identify yours.

From now on in this book, we'll abbreviate menu selections with the name of the menu and the submenu or command. The command you just gave would be **Graphs ➤ Bar...**

🖱 **Insert ➤ New Page Title** On the output pane, a box will appear. Type your name in it, and then move the cursor just outside of the box.[3] Click the left mouse button.

Before evaluating the bar chart, look at the bottom of your screen. There are two buttons with SPSS icons on them. One says swim – SPSS for Windows..., and the other Output1 – SPSS for Windows... You can switch between the Data Editor and the Viewer window by clicking on one button or the other.

Now look at the chart. The height of each bar corresponds to the simple average time of the males and females. ***What does the chart tell you about the original question: did the males or females have a better practice that day?***

There is more to a set of data than its average. Let's look at another graph that can give us a feel for how the swimmers did individually and collectively. This graph is called a *box-and-whiskers* plot (or *boxplot*), and displays how the swimmers' times were spread out. Boxplots are fully discussed in Session 4, but we'll take a first look now. You may issue this command either from the Data Editor or the Viewer.

🖱 **Graphs ➤ Boxplot...** As with the bar chart, you get a small dialog box asking for the type of boxplot we want. We'll use the default, and just click **Define**.

🖱 Complete this dialog box as shown to the right, and click **OK**.

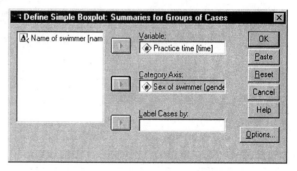

This command generates two outputs, as shown on the facing page. The Case Processing Summary is found in many SPSS commands. It lets us know how many of the cases in the sample have valid values for our variable (Time). Here, we have no missing data.

The Case Processing Summary is followed by the boxplot, which shows results for the males and females. There are two boxes, and each

---

[3] If you click near the graph itself, you will select the graph and a box will appear around it. Later you'll learn about selecting parts of your output. To deselect the graph, click anywhere outside of the selected box.

has *whiskers* extending above and below the box. In this case, the whiskers extended from the shortest to the longest time. The outline of the box reflects the middle three times, and the line through the middle of the box represents the *median* value for the swimmers.[4]

Explore
Sex of swimmer

Case Processing Summary

		Cases					
		Valid		Missing		Total	
	Sex of swimmer	N	Percent	N	Percent	N	Percent
Practice time	Female	5	100.0%	0	.0%	5	100.0%
	Male	5	100.0%	0	.0%	5	100.0%

Practice time

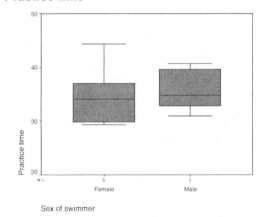

Sex of swimmer

Looking now at the boxplot, what impression do you have of the practice times for the male and female swimmers? How does this compare to your impression from the first graph?

Saving an Output File

At this point, we have the Viewer window open with some output and the Data Editor with a data file. We have saved the data, but have not yet saved the output on a disk.

---

[4] The median of a set of points is the middle value when the observations are ranked from smallest to largest.

File ➤ Save As... In this dialog box, assign a name to the file (such as Session 1). This new file will save both the Outline and Content panes of the Viewer window.

Getting Help

You may have noticed the **Help** button in the dialog boxes. SPSS features an extensive on-line Help system. If you aren't sure what a term in the dialog box means, or how to interpret the results of a command, click on Help. You can also search for help on a variety of topics via the Help menu at the top of your screen. As you work your way through the sessions in this book, Help may often be valuable. Spend some time experimenting with it before you genuinely need it.

Printing in SPSS

Now that you have created some graphs, let's print them. Be sure that no part of the Outline is highlighted; if it is, click once in a clear area of the Outline pane. If a portion of the outline is selected, only that portion will print.

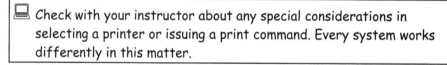 Check with your instructor about any special considerations in selecting a printer or issuing a print command. Every system works differently in this matter.

File ➤ Print... This command will print the output window. Click **OK**.

Quitting SPSS

When you have completed your work, it is important to exit the program properly. Virtually all Windows programs follow the same method of quitting.

File ➤ Exit You will generally see a message asking if you wish to save changes. Since we saved everything earlier, click **No**.

That's all there is to it. Later sessions will explain menus and commands in greater detail. This session is intended as a first look; you will return to these commands and others at a later time.

Index